Etale Homotopy
of Simplicial Schemes

INTRODUCTION

In the early part of this century, algebraic geometry and algebraic topology were not separate disciplines. Indeed many manifolds admit the structure of a (complex) algebraic variety. S. Lefschetz and others introduced algebro-geometric techniques to study topological properties of such varieties. Subsequently, algebraic geometry widened its scope to include varieties of positive characteristic, rings arising in algebraic number theory, and more general schemes. An algebrization of algebraic geometry has been achieved, thanks in large part to A. Weil and O. Zariski; and powerful techniques of homological algebra have been employed, especially those introduced by J.-P. Serre and A. Grothendieck. On the other hand, algebraic topology has developed in directions not obviously relevant to algebraic geometry.

Inspired by A. Weil's celebrated conjectures, M. Artin and A. Grothendieck developed etale cohomology theory, a theory sufficiently well behaved that in some ways it plays the role for positive characteristic varieties of singular cohomology theory in algebraic topology. P. Deligne's proof of the Weil Conjectures [22] is a dramatic application of etale cohomology. M. Artin and B. Mazur refined etale cohomology theory in their book entitled *Etale Homotopy* [8]: they introduced a "pro-homotopy type" associated to a scheme whose cohomology is the etale cohomology of the scheme. Thus, etale homotopy theory reinforces the relationship between algebraic geometry and algebraic topology.

The purpose of this book is to provide a coherent account of the current state of etale homotopy theory. We describe in Chapters 9, 11, and 12, various applications of this theory to algebraic topology, cohomology of groups, and algebraic geometry. These applications have required repeated refinement and generalization of the theory developed by M. Artin and B. Mazur. For this reason, we consider simplicial schemes throughout, necessary for almost all existing applications, and introduce the etale topological type which refines the etale homotopy type.

With an eye to future applications, we introduce various algebraic invariants of the etale topological type not usually considered by algebraic geometers: function complexes, relative homology and cohomology, and generalized cohomology. It is our hope that this presentation of etale homotopy theory will enable applications by topologists and geometers who have not immersed themselves in the delicate technicalities of the subject. Consequently, we have attempted to minimize pre-requisites to basic homological algebra [17], elementary theory of simplicial sets [57], and an acquaintance with algebraic geometry [50]. We have taken this opportunity to re-work much of the foundational material of etale homotopy developed by M. Artin-B. Mazur and the author: the knowledgeable reader will recognize many improvements and generalizations of results in the literature.

We now proceed to briefly sketch the contents and organization of the various chapters, each of which also has its own introduction. Beginning with the definition of a simplicial scheme, Chapter 1 describes examples most common in applications and briefly discusses a technique for adapting constructions on schemes to apply to simplicial schemes. Chapter 2 defines (etale) sheaf cohomology in terms of derived functors and relates the sheaf cohomology of a simplicial scheme to that of its constituent schemes. After discussing the somewhat more familiar Čech construction, we describe in Chapter 3 the analogous construction of J.-L. Verdier which enables one to compute cohomology in terms of hypercoverings. Chapter 4 presents the etale topological type of a simplicial scheme based on the inverse system of rigid hypercoverings (initially inspired by work of S. Lubkin [54]). This etale topological type is a refinement of the Artin-Mazur etale homotopy type and is given by a somewhat more natural construction than previous refinements. In Chapter 5, we verify that the set of connected components, the fundamental group, and the cohomology with local coefficients of the etale topological type of a simplicial scheme are given by the set of connected components, the Grothendieck fundamental group, and the sheaf cohomology of the simplicial scheme.

Because the etale topological type is not a single simplicial set but an inverse system, machinery must be employed to analyze its homotopy invariants. Chapter 6 compares constructions of M. Artin-B. Mazur, D. Sullivan, and A. K. Bousfield-D. M. Kan, each of which has been employed in the literature for various applications. To assist in the identification of various homotopy invariants using these constructions, we prove several finiteness properties in Chapter 7. Chapter 8 then relates the etale topological type of a simplicial complex variety to the geometric realization of its associated simplicial (analytic) space and provides a comparison of topological types in characteristic $p \neq 0$ and 0. An especially important example is that of the classifying space of a complex reductive Lie group.

Having travelled so far, the reader is rewarded with various applications. Interest in etale homotopy theory was aroused by D. Quillen's use of the theory in his sketch of a proof of the (complex) Adams conjecture [61]. We present a modification of D. Sullivan's proof of the Adams conjecture [69] as well as our proof of its infinite loop space refinement [37]. Chapter 9 also describes the author's use of homomorphisms of algebraic groups in positive characteristic to provide interesting maps of classifying spaces and homogeneous spaces of compact Lie groups [33], [36]. Chapter 11 presents four geometric applications of etale homotopy theory originally due to Deligne-Sullivan [25], D. Cox [20], and the author [30], [31]. The last two applications of Chapter 11 and all those of Chapter 12 require a comparison of homotopy types of the algebro-geometric and homotopy fibres as discussed in Chapter 10. In Chapter 12, we describe further applications by the author of etale homotopy to K-theories of finite fields and the cohomology of finite Chevalley groups [32], [34], [35], [41].

The discussion of function complexes in Chapter 13 is central to recent applications of etale homotopy to algebraic K-theory by the author and W. Dwyer [27], [39]. Because this is the first published account of function complexes, Chapter 13 examines their behavior in some detail. In Chapter 14, we discuss relative cohomology in order to incorporate local cohomology and cohomology with supports in the framework of etale homotopy theory. Chapter 15 provides a refinement of D. Cox's construction of tubular neighborhoods [18] and utilizes these tubular neighborhoods in a discussion of excision and Mayer-Vietoris. Chapter 16 provides a first introduction of generalized cohomology into the study of algebraic geometry: an early version of this chapter was the basis for the author's comparison of algebraic and topological K-theory of varieties [38]. Finally, Chapter 17 presents a sketch of Poincaré duality for etale cohomology using the machinery developed in the preceding chapters.

During the lengthy evolution of this book, further developments and applications of etale homotopy theory have arisen. We refer the reader to [73] for a discussion of sheaves of spectra and etale cohomological descent utilized in a further application to algebraic K-theory.

We are especially indebted to D. Cox, who shared with us pre-prints of his work on the homotopy type of simplicial schemes [19] and on tubular neighborhoods [18]. We also thank A. K. Bousfield, W. Dwyer, and R. Thomason for many valuable conversations. Finally, we thank Oxford University, the University of Cambridge, and the Institute for Advanced Study for their warm hospitality during the writing of this book, and gratefully acknowledge support from the National Science Foundation and the Scientific Research Council.

1. ETALE SITE OF A SIMPLICIAL SCHEME

After stating the definition of a simplicial scheme, we provide several important examples in Examples 1.1, 1.2, and 1.3. The first two examples arise frequently in applications, whereas the third is central in the construction of the etale homotopy type. In Definition 1.4, we introduce the etale site of a simplicial scheme. As we explain, the etale site is a generalization due to Grothendieck of the category of open subsets of a topological space. We conclude this chapter with a construction which enables one to extend certain arguments applied to a given dimension of a simplicial scheme to the entire simplicial scheme.

We recall that Δ, the *category of standard simplices*, consists of objects $\Delta[n]$ for $n \geq 0$ and maps $\Delta[n] \to \Delta[m]$ for each nondecreasing function from $\{0,1,\cdots,n\}$ to $\{0,1,\cdots,m\}$. Any such map $\alpha : \Delta[n] \to \Delta[m]$ (other than the identity) can be written as a composite of "degeneracy" maps $\sigma_j : \Delta[k] \to \Delta[k-1]$ with $j \leq k-1$ defined by $\sigma_j(j) = \sigma_j(j+1) = j$ and σ surjective and "face" maps $\partial_i : \Delta[\ell] \to \Delta[\ell+1]$ with $i \leq \ell+1$ defined as that injective map such that $i \in \{0,\cdots,\ell+1\}$ is not in the image of ∂_i. A *simplicial object* of a category C is a functor $\Delta^0 \to C$, where Δ^0 is the opposite category to Δ; a map of simplicial objects is a natural transformation of functors.

A *simplicial scheme* is a functor from Δ^0 to the category of schemes (= local ringed spaces which admit a covering by open subspaces isomorphic to the spectra of rings; cf., [50], II.2). Following conventional notation, we denote such a simplicial scheme by $X.$, and we denote $X.(\Delta[n])$ by X_n.

EXAMPLE 1.1. Let X be a scheme and $T.$ a simplicial set. Then $X \otimes T.$ is the simplicial scheme with $(X \otimes T.)_n$ equal to the disjoint union of copies of X indexed by T_n and with $\alpha : (X \otimes T.)_m \to (X \otimes T.)_n$ for $\alpha : \Delta[n] \to \Delta[m]$ in Δ given by sending the summand of X indexed by $t \in T_m$ via the identity to the summand of X indexed by $\alpha(t) \in T_n$. More generally, if $X.$ is a simplicial scheme and $T.$ a simplicial set, then $X. \otimes T.$ is similarly defined with $(X. \otimes T.)_n$ equal to the disjoint union of copies of X_n indexed by T_n.

Let $\Delta[k]$ also denote the simplicial set defined by $\Delta[k]_n = \mathrm{Hom}_\Delta(\Delta[n], \Delta[k])$. In particular, $X \otimes \Delta[0]$ is the simplicial scheme which is equal to X in each dimension (we shall usually denote $X \otimes \Delta[0]$ simply by X). We easily verify that a map of simplicial schemes $X \otimes \Delta[k] \to Y.$ is equivalent to a map of schemes $X \to Y_k$. A special case of Example 1.1 which we shall frequently employ is that in which $T.$ equals $\Delta[1]$ so that $(X. \otimes \Delta[1])_n$ is the disjoint union of $n+1$ copies of X_n. Two maps $f,g : X. \to Y.$ of simplicial schemes are said to be related by the *simplicial homotopy* $H : X. \otimes \Delta[1] \to Y.$ if the two canonical restrictions of H to $X. \otimes \Delta[0]$ are f and g.

If X is a scheme which appears acyclic with respect to a certain cohomology theory (e.g., X might be the spectrum of an algebraically closed field), then $X \otimes T.$ appears to be an algebraic model of the simplicial set $T.$ for this cohomology theory. This is exploited in Chapter 11 to study finite Chevalley groups, as well as by Z. Wojtkowiak in [72], and (implicitly) by C. Soulé in [67]. Using the special case $T = B\mathbb{Z}/2$, D. Cox has studied real algebraic varieties (see Corollary 11.3). ∎

EXAMPLE 1.2. Let S be a scheme. A *group scheme over* S, G, is a map of schemes $G \to S$ with the property that $\mathrm{Hom}_S(\ ,G)$ is a group valued functor on the category of schemes over S (i.e., maps of schemes $Z \to S$). In other words, G is provided with maps $e : S \to G$ and $\mu : G \times_S G \to G$ over S satisfying the usual axioms. If G is a group scheme over S, the *classifying simplicial scheme* BG is the simplicial scheme given in dimension $n \geq 0$ by $(BG)_n = G^{\times n}$, the n-fold fibre product of G with itself over S, and with structure maps given in the usual way using e and μ. More generally, if X (respectively, Y) is a scheme over S provided with left G-action $\alpha: G \times_S X \to X$ (resp., right G-action $\beta: Y \times_S G \to G$), then $B(Y,G,X)$ is the simplicial scheme given in dimension $n \geq 0$ by

$$B(Y,G,X)_n = Y \times_S G^{\times n} \times_S X$$

and with structure maps given in the usual way using e, μ, α and β (cf., [58], for an explicit description of this "*double bar construction*"). We let $B(Y,G,*)$ (resp., $B(*,G,X)$) be the simplicial schemes obtained from $B(Y,G,X)$ by deleting the right-hand (resp., left-hand) factor.

Special cases of Example 1.2 have occurred in most of the applications of etale homotopy theory. These examples are so prevalent because any complex reductive Lie group $G(\mathbb{C})$ admits the structure of an algebraic group $G_{\mathbb{C}}$ over Spec $\mathbb{C}$; $BG_{\mathbb{C}}$ then provides an algebraic model for the topological classifying space of $G(\mathbb{C})$. Moreover, if H is a subgroup scheme of G over S, then H acts on G by multiplication so that $B(G,H,*)$ and $B(*,H,G)$ serve as models for the "homogeneous space" G/H (see Chapter 9). ■

EXAMPLE 1.3. Let $X.$ be a simplicial scheme over a given scheme S. For $n \geq 0$, let $X.^{(n)}$ denote the n-th truncation of $X.$: $X.^{(n)}$ is the restriction of $X.$ to the full-subcategory of Δ^0 with objects $\Delta[k]$, $k \leq n$. For any $n \geq 0$, we define the n-th *coskeleton* of $X.$ over S, $\operatorname{cosk}_n^S X.$ (or $\operatorname{cosk}_n X.$ if no confusion arises by leaving S implicit) by the following universal property: For any simplicial scheme $Y.$ over S, the set of maps $Y. \to \operatorname{cosk}_n^S X.$ over S is in natural one-to-one correspondence with the set of maps $Y.^{(n)} \to X.^{(n)}$ over S. If we view $\operatorname{cosk}_n^S(\ )$ as a functor from n-truncated simplicial schemes to simplicial schemes, then $\operatorname{cosk}_n^S(\ )$ is right adjoint to the n-truncation functor $(\ )^{(n)}$.

If X is a scheme over S, $\operatorname{cosk}_0^S X$ ($= \operatorname{cosk}_0^S(X \otimes \Delta[0])$) is the *Čech nerve* (usually written $N_S(X)$) of X over S with $(\operatorname{cosk}_0^S X)_n$ equal to the (n+1)-fold fibre product of X with itself over S. One verifies that

$\operatorname{cosk}_n^S X.$ is, in general, a finite fibre product involving the X_k's with $k \leq n$ and S. For example,

$$(\operatorname{cosk}_1^S X.)_2 = (X_1 \times_{X_0} X_1) \times_{(X_0 \times_S X_0)} X_1 \,.$$

As we shall see in Chapter 3, Čech nerves of "etale open coverings" and more general etale hypercoverings (defined in terms of coskeletons) will be used to compute cohomology and define the etale homotopy type. ∎

We next turn to the definition of the etale site of a simplicial scheme $X.$: the category of etale opens of $X.$ and the Grothendieck topology we impose upon them (see Definition 1.4 below). Because every scheme has an underlying (Zariski) topological space, we could view a simplicial scheme as a simplicial topological space by forgetting the sheaf of rings on each scheme. From a homotopy theoretic point of view, this is uninteresting: If A is an integral domain, then any two Zariski open subsets of Spec A intersect so that the nerve of any Zariski covering yields a contractible simplicial set of connected components.

In introducing the etale "topology", Grothendieck had the insight to see that sheaf theory on a topological space could be employed on more general "topologies." In the etale topology, the "open sets" are more general than Zariski open sets: Any *etale map* (i.e., a disjoint union of finitely presented, flat, and unramified maps) $U \to X$ is considered an "etale open." The reader should be wary that an etale open $U \to X$ need not be an inclusion and that there may be nontrivial self maps of an etale open.

The critical observation that suggests that there are enough etale opens of a scheme for certain purposes is the following theorem of Zariski and Artin (see Theorem 11.6). Namely, every smooth point on a complex algebraic variety admits a Zariski open neighborhood whose universal covering space is contractible and whose fundamental group is a successive extension of free groups. This implies that any positive degree cohomology class with finite coefficients on an algebraic variety vanishes when restricted to sufficiently fine etale neighborhoods of a smooth point, because a finite covering space of a Zariski open set is an etale open.

DEFINITION 1.4. Let $X.$ be a simplicial scheme. Let $Et(X.)$ denote the category whose objects are etale maps $U \to X_n$ for some $n \geq 0$, and whose maps are commutative squares

$$\begin{array}{ccc} U & \longrightarrow & V \\ \downarrow & & \downarrow \\ X_n & \longrightarrow & X_m \end{array}$$

whose bottom arrow is a specified structure map of $X.$ (i.e., a choice of map $\alpha : \Delta[m] \to \Delta[n]$ is part of the data of a map in $Et(X.)$). Give $Et(X.)$ the *etale topology* with a *covering* of a given object $U \to X_n$ defined to be a family of etale maps $\{U_i \to U\}_{i \in I}$ over X_n with the union of the images of U_i in U covering U. Let $Et(X.)$ also denote the resultant *etale site* consisting of the category $Et(X.)$ together with this etale topology. ∎

Because the composition of etale maps is etale, we may view $U_i \to U$ as a map of $Et(X.)$. At times, we shall identify $\{U_i \to U\}_{i \in I}$ with the single etale surjective map $\coprod_{i \in I} U_i \to U$. (It is at this point that we employ the convention that an arbitrary disjoint union of finitely presented, flat, and unramified maps be etale.)

Let $X.$ be a simplicial scheme, and let $U \to X_n$ be an object of $Et(X.)$. There does not necessarily exist a map of simplicial schemes, $U. \to X.$, given in dimension n by $U \to X_n$. Nonetheless, the following proposition enables us to extend constructions on a given dimension of a simplicial scheme to global constructions on the entire simplicial scheme.

PROPOSITION 1.5. *Let* $X.$ *be a simplicial scheme. For any* $n \geq 0$, *the restriction functor*

$$(\ )_n : (\text{s. schemes}/X.) \to (\text{schemes}/X_n)$$

sending $Z. \to X.$ *to* $Z_n \to X_n$ *admits a right adjoint*

$$\Gamma_n^{X.}(\ ) : (\text{schemes}/X_n) \to (\text{s. schemes}/X.) .$$

Moreover, if $W \to X_n$ *is etale (respectively, surjective), then* $\Gamma_n^{X.}(W) \to X.$ *is also etale (resp., surjective) in each dimension.*

Proof. We recall that a functor $G : D \to C$ is said to be right *adjoint* to a functor $F : C \to D$ if there exist natural bijections $\mathrm{Hom}_C(X,G(Y)) \xrightarrow{\sim} \mathrm{Hom}_D(F(X),Y)$ for $X \in C$, $Y \in D$ (cf., [55]). Let $W \to X_n$ be given. We define $\Gamma_n^{X.}(W)_k$ to be the fibre product of $W^{\times \Delta[k,n]} \to X_n^{\times \Delta[k,n]} \leftarrow X_k$, where $\Delta[k,n] = \Delta[k]_n$ and where X_k maps into that factor of X_n indexed by $\alpha : \Delta[n] \to \Delta[k]$ by $\alpha : X_k \to X_n$. The adjointness of $\Gamma_n^{X.}(\ )$ is readily checked, whereas the last assertion follows from the facts that the product and pullback of etale (resp., surjective) maps are again etale (resp., surjective). ∎

The canonical map $\Gamma_n^{X.}(W) \to X.$ associated to $W \to X_n$ is easily seen to have the property that its restriction to dimension n, $(\Gamma_n^{X.}(W))_n \to X_n$, factors through $W \to X_n$. Thus, as an immediate corollary of Proposition 1.5, we conclude that every etale surjective map $W \to X_n$ is dominated by the restriction to dimension n of some etale surjective map $U. \to X.$.

2. SHEAVES AND COHOMOLOGY

In this chapter, we consider sheaves on the etale site $Et(X.)$ of a simplicial scheme. As defined in Definition 2.3, the cohomology groups of abelian sheaves are derived functors (in fact, Ext functors). Proposition 2.4 provides the relationship between sheaf cohomology of a simplicial scheme $X.$ and that of X_n for $n \geq 0$. We conclude this chapter with a brief discussion of sheaf cohomology of bi-simplicial schemes, showing in Proposition 2.5 how this reduces to sheaf cohomology of simplicial schemes.

Before we give the definition of a sheaf, we introduce the concept of a presheaf on a simplicial scheme $X.$. Namely, an (etale) *presheaf* of sets on $X.$ is a functor $P: Et(X.)^0 \to (\text{sets})$ (where $Et(X.)^0$ is the opposite category of $Et(X.)$ defined in Definition 1.4). For notational convenience, if P is a presheaf on $X.$, and $U \to X_n$ is an object of $X.$, we write $P(U)$ for the value of P on $U \to X_n$.

DEFINITION 2.1. Let $X.$ be a simplicial scheme. Then a *sheaf* of sets on $Et(X.)$ is a functor $F: Et(X.)^0 \to (\text{sets})$ satisfying the following *sheaf axiom*: For every covering $\{U_i \to U\}_{i \in I}$ in $Et(X.)$, the set $F(U)$ is the equalizer of the two natural maps

$$\prod_{i \in I} F(U_i) \rightrightarrows \prod_{\langle i,j\rangle \in I \times I} F(U_i \times_U U_j) . \blacksquare$$

In other words, such a sheaf F consists of the data of sheaves F_n on $Et(X_n)$ (the etale site of X_n) for each $n \geq 0$ together with maps $F_n \to \alpha_* F_m$ of sheaves on $Et(X_n)$ for each $\alpha : \Delta[n] \to \Delta[m]$ in Δ, where $\alpha_* F_m(U) = F_m(U \times_{X_n} X_m)$. We remind the reader that α_* has a left adjoint α^* ([59], II.2.2), so that the sheaf data is equivalent to sheaves F_n on

$Et(X_n)$ and maps $\alpha^*F_n \to F_m$. In order to determine a sheaf on Et(X.), this data must satisfy the compatibility condition that the composition $\alpha^*(\beta^*F_p) \to \alpha^*F_n \to F_m$ equals the map $(\beta\circ\alpha)^*F_p \to F_m$ associated with $\alpha\circ\beta : \Delta[p] \to \Delta[n] \to \Delta[m]$. In particular, if $X. = X \otimes \Delta[0]$, then a sheaf on Et(X.) is equivalent to a cosimplicial object of sheaves on Et(X).

If W is a scheme, then W determines the *representable* (in the large) *sheaf* of sets defined by sending $U \to X_n$ in Et(X.) to the set $\mathrm{Hom}_{(\mathrm{schemes})}(U,W)$ of all morphisms of schemes from U to W. Similarly, $V \to X_m$ in Et(X.) determines the representable (in the small) sheaf of sets defined by sending $U \to X_n$ to $\mathrm{Hom}_{Et(X.)}(U,V)$. The *constant sheaf* C (for any set C) is the sheaf represented by $\mathrm{Spec}\, \mathbf{Z} \otimes C$, the scheme given as a disjoint union of copies of $\mathrm{Spec}\,\mathbf{Z}$ indexed by the set C. A *locally constant sheaf* F on Et(X.) is a sheaf on Et(X.) such that F_0 restricted to each $U_i \to X_0$ of some covering $\{U_i \to X_0\}$ in Et(X.) is constant, and such that $\alpha^*F_n \to F_m$ is an isomorphism for all $\alpha : \Delta[n] \to \Delta[m]$. For example, a representation of the fundamental group of the simplicial set T. pointed by $t \in T_0$ in the automorphisms of an abelian group A, $\pi_1(T.\,,t) \to \mathrm{Aut}(A)$, determines a locally constant sheaf of abelian groups on $Et(X. \otimes T.)$ for any simplicial scheme X.. Such a sheaf is not constant, even though its restriction to $X_0 \otimes T_0$ is constant.

We briefly consider the special case of a field k. In this case, any $U \to \mathrm{Spec}\, k$ in Et(Spec k) is dominated by a finite, galois field extension $\mathrm{Spec}\, K \to \mathrm{Spec}\, k$, and a sheaf F of abelian groups on Et(Spec k) is equivalent to the data of the group $\operatorname{colim}_{K/k} F(\mathrm{Spec}\, K \to \mathrm{Spec}\, k) = F(\bar{k})$ provided with a left action of the galois group $\varprojlim_{K/k} \mathrm{Gal}(K/k)$ such that the stabilizer of any element of $F(\bar{k})$ is a subgroup of finite index ([59], II.1.9). In particular, if k is a separably (algebraically) closed field, than a sheaf F on Et(Spec k) is equivalent to its value F(Spec k). Viewing such Spec k as "points" for the etale topology, we define a *geometric point* of a simplicial scheme X. to be a map $\mathrm{Spec}\, k \to X_m$, some $m \geq 0$

and some separably closed field k. We define the *stalk* of a sheaf F on $Et(X.)$ at the geometric point $\alpha: \operatorname{Spec} k \to X_m$ to be $\alpha^* F_m(\operatorname{Spec} k)$.

We proceed to the definition of the (etale) cohomology of an *abelian sheaf* (i.e., a sheaf of abelian groups) F on $Et(X.)$ using derived functors. For a more detailed discussion, we refer the reader to [59].

PROPOSITION 2.2. *Let* $X.$ *be a simplicial scheme, and let* $AbSh(X.)$ *denote the category of abelian sheaves on* $Et(X.)$. *Then* $AbSh(X.)$ *is an abelian category with enough injectives. Moreover, a sequence of sheaves in* $AbSh(X.)$ *is exact if and only if for every geometric point* $\alpha: \operatorname{Spec} k \to X_m$ *of* $X.$ *the sequence of stalks at* α *is exact.*

Proof. A map $A \to B$ in $AbSh(X.)$ has kernel $K \to A$ (respectively, cokernel $B \to C$) if and only if the restriction $K_n \to A_n$ (resp., $B_n \to C_n$) is the kernel (resp., cokernel) of $A_n \to B_n$ in $AbSh(X_n)$ for each $n \geq 0$, where $AbSh(X_n)$ is the abelian category of abelian sheaves on $Et(X_n)$ (cf. [59], II.2.15). This readily implies that $AbSh(X.)$ is an abelian category. To show $AbSh(X.)$ has enough injectives, we use the functor $R_n(\ ): AbSh(X_n) \to AbSh(X.)$ defined by sending $G \in AbSh(X_n)$ to $R_n(G)$ with $(R_n(G))_m = \prod_{\Delta[n]_m} \alpha_* G$. Because $R_n(\ )$ is right adjoint to the restriction functor (which is exact), $R_n(G)$ is injective in $AbSh(X.)$ whenever G is injective in $AbSh(X_n)$. Thus, if F is an arbitrary abelian sheaf on $Et(X.)$ and if $F_n \to I_n$ is a monomorphism of F_n in an injective I_n of $AbSh(X_n)$ for each $n \geq 0$, then $F \to \prod_{n=0}^{\infty} R_n(I_n)$ is a monomorphism of F into an injective of $AbSh(X.)$. Because a sequence of sheaves in $AbSh(X.)$ is exact if and only if its restriction to each $AbSh(X_n)$ is exact, the last statement follows from the corresponding fact for $AbSh(X_n)$ (cf. [59], II.2.15). ∎

The following definition of the (etale) cohomology of a simplicial scheme (due to P. Deligne in [23], 5.2) is the natural generalization of the definition of cohomology of a scheme as derived functors of the global section functor.

DEFINITION 2.3. Let X. be a simplicial scheme. For any $i \geq 0$, the *cohomology group* functor

$$H^i(X.\,,\ ): AbSh(X.) \to Ab$$

is the i-th right derived functor of the functor sending an abelian sheaf F on Et(X.) to the abelian group given as the kernel of the map $d_0^* - d_1^* : F(X_0) \to F(X_1)$. Equivalently,

$$H^i(X.\,,\ ) = Ext^i_{AbSh(X.)}(\mathbf{Z},\ )$$

where $\mathbf{Z}$ is the constant abelian sheaf on Et(X.) with each stalk equal to $\mathbf{Z}$. ■

The definition of $H^*(X.\,,F)$ is global in the sense that it is given in terms of Et(X.) rather than the various $Et(X_n)$. Nonetheless, the following proposition enables us to relate $H^*(X.\,,F)$ to the various $H^*(X_n, F_n)$, $n \geq 0$.

PROPOSITION 2.4. *Let* X. *be a simplicial scheme, and let* $F \in AbSh(X.)$. *Then there exists a first quadrant spectral sequence*

$$E_1^{s,t} = H^t(X_s, F_s) \Rightarrow H^{s+t}(X.\,,F)$$

natural with respect to X. *and* F.

Proof. Let $F \to I^{\cdot}$ be an injective resolution in AbSh(X.). We consider the functor $L^n(\ ): AbSh(X_n) \to AbSh(X.)$ defined by sending $G \in AbSh(X_n)$ to $L^n(G)$ defined by $(L^n(G))_m = \bigoplus_{\alpha \in \Delta[m]_n} \alpha^* G$, any $n \geq 0$. Because $L^n(\ )$ is an exact left adjoint to the restriction functor $(\ )_n$, we conclude that $(\ )_n$ sends injectives to injectives for each $n \geq 0$. Thus, $F_n \to (I^{\cdot})_n$ is an injective resolution in $AbSh(X_n)$.

Let $\mathbb{Z}_{X_*} = \mathbb{Z}_*$ denote the complex of abelian sheaves in AbSh(X.) whose m-th term is the abelian sheaf represented (in the small) by $id : X_m \to X_m$ in $Et(X_m)$ (and whose differential is the usual alternating sum of the maps induced by $d_i : X_m \to X_{m-1}, i \leq m$). Thus, $\mathbb{Z}_m$ equals $L^m(\mathbb{Z}_{|X_m})$; when restricted to $Et(X_n)$, $\mathbb{Z}_m$ is the direct sum indexed by $\Delta[n]_m$ of copies of the constant sheaf $\mathbb{Z}$. We define an augmentation map $\mathbb{Z}_{X_*} \to \mathbb{Z}$ by defining $\mathbb{Z}_0(U) = \underset{\Delta[n]_0}{\oplus} \mathbb{Z}(U) \to \mathbb{Z}(U)$ for any $U \to X_n$ in Et(X.) to be the identity on each summand. To verify that $\mathbb{Z}_{X_*} \to \mathbb{Z} \to 0$ is exact in AbSh(X.), it suffices to verify that the restrictions $(\mathbb{Z}_{X_*})_n \to \mathbb{Z} \to 0$ are exact in $AbSh(X_n)$ for $n \geq 0$. This exactness follows from the acyclicity of $\Delta[n]$ and the identification of $(\mathbb{Z}_{X_*})_n \to \mathbb{Z}$ with $\mathbb{Z} \otimes (\Delta[n] \to \Delta[0])$.

The spectral sequence is obtained from the bi-complex $Hom_{AbSh(X.)}(\mathbb{Z}_{X_*}, I^{\cdot})$: the total cohomology of the bi-complex is that of $Hom_{AbSh(X.)}(\mathbb{Z}, I^{\cdot})$ which is $H^*(X.\,,F)$; whereas the cohomology of $Hom_{AbSh(X.)}(\mathbb{Z}_m, I^{\cdot}) = Hom_{AbSh(X_m)}(\mathbb{Z}, (I^{\cdot})_m)$ is $H^*(X_m, F^m)$. ■

As an immediate corollary of Proposition 2.4, we conclude that if $F \in AbSh(X.)$ is such that $F_n \in AbSh(X_n)$ is injective for each $n \geq 0$, then F is *acyclic* (i.e., $H^i(X.\,,F) = 0$ for $i > 0$) if and only if the cochain complex $F(X.) \equiv \{n \mapsto F(X_n)\}$ is acyclic.

We shall often utilize bi-simplicial schemes to study simplicial schemes. As the reader can see from the following brief discussion, the etale topology and etale cohomology of bi-simplicial schemes are defined analogously to that of simplicial schemes. Moreover, Proposition 2.5 permits us to replace a bi-simplicial scheme X.. by its *diagonal simplicial scheme* ΔX.. (where $(\Delta X..)_n = X_{n,n}$).

A *bi-simplicial scheme* is a functor from $\Delta^0 \times \Delta^0$ to the category of schemes. Following convention, we denote such a bi-simplicial scheme by X.. , and we denote $X..(\Delta[m], \Delta[n])$ by $X_{m,n}$. We let Et(X..) denote the category whose objects are etale maps $U \to X_{s,t}$ for some $s,t \geq 0$ and whose maps are commutative squares

$$\begin{array}{ccc} U & \longrightarrow & V \\ \big\downarrow & & \big\downarrow \\ X_{s,t} & \longrightarrow & X_{k,\ell} \end{array}$$

whose bottom arrow is a specified structure map of $X..$. We give Et(X..) the etale topology by defining a covering of $U \to X_{s,t}$ to be a family of etale maps $\{U_i \to U\}_{i \in I}$ over $X_{s,t}$ whose images cover U. We let Et(X..) also denote the resultant etale site on the category Et(X..).

We define AbSh(X..) to be the abelian category of abelian sheaves on the site Et(X..), where the sheaf axiom is that of Definition 2.1. The (etale) *cohomology groups of a bi-simplicial scheme* X.. with values in a sheaf $F \in$ AbSh(X..) are defined to be

$$H^i(X.\,,F) = \mathrm{Ext}^i_{\mathrm{Absh}(X..)}(\mathbf{Z},F)$$

where $\mathbf{Z}$ is the constant abelian sheaf on Et(X.) with fibres $\mathbf{Z}$.

PROPOSITION 2.5. *Let* X.. *be a bi-simplicial scheme. Then there is a natural isomorphism of δ-functors*

$$H^*(X..\,,\;) \xrightarrow{\simeq} H^*(\Delta X..\,,(\;)^\Delta)$$

on AbSh(X..), *where* $\Delta X..$ *is the diagonal (simplicial scheme) of* X.., *and* $(\;)^\Delta : \mathrm{AbSh}(X..) \to \mathrm{AbSh}(\Delta X..)$ *is the restriction functor.*

Proof. It suffices to (a) exhibit a natural isomorphism $H^0(X..\,,\;) \xrightarrow{\simeq} H^0(\Delta X..\,,(\;)^\Delta)$, (b) prove $H^i(\Delta X..\,,I^\Delta) = 0$ for $i > 0$ any $I \in$ AbSh(X..) injective, and (c) verify that $H^*(\Delta X..\,,(\;)^\Delta)$ is in fact a δ-*functor* (i.e., there is naturally associated a long exact sequence

$$\cdots \to H^i(\Delta X.. ,F^\Delta) \to H^i(\Delta X.. ,G^\Delta) \to H^i(\Delta X.. ,H^\Delta) \to H^{i+1}(\Delta X.. ,F^\Delta) \to \cdots$$

to each short exact sequence $0 \to F \to G \to H \to 0$ in AbSh(X..)) (cf., [17], V. 4.4). Part (a) is readily achieved by identifying both $H^0(X.., \ )$ and $H^0(\Delta X.. ,(\ )^\Delta)$ with the group $\mathrm{Ker}(d_0^* \circ d_0^* - d_1^* \circ d_1^* : F(X_{0,0}) \to F(X_{1,1}))$. Arguing as in the proof of Proposition 2.4, we verify that $I_{s,t} \in \mathrm{AbSh}(X_{s,t})$ is injective whenever $I \in \mathrm{AbSh}(X..)$ is injective. Thus, to prove (b), it suffices to prove that $I^\Delta(\Delta X..) \equiv \{n \mapsto I(X_{n,n})\}$ is acyclic whenever $I \in \mathrm{AbSh}(X..)$ is injective. Because Tot(I(X..)) and $\Delta I(X..) = I^\Delta(\Delta X..)$ have the same cohomology ([26], 2.9), this follows from the exactness of $\mathrm{Tot}(Z_{X..})$ whose proof is analogous to that of the exactness of $Z_{X.}$ given in Proposition 2.4. Finally, to prove (c), it suffices to observe that $(\ )^\Delta$ is exact because $0 \to F \to G \to H \to 0$ is exact in AbSh(X..) if and only if $0 \to F_{s,t} \to G_{s,t} \to H_{s,t} \to 0$ is exact in $\mathrm{AbSh}(X_{s,t})$ for all $s,t \geq 0$. ∎

We conclude this chapter with the following bi-simplicial analogue of Proposition 2.4.

PROPOSITION 2.6. *Let* X.. *be a bi-simplicial scheme and let* F *be an abelian sheaf on* Et(X..). *Then there exists a first quadrant spectral sequence*

$$E_1^{s,t} = H^t(X_{s.} ,F_{s.}) \Longrightarrow H^{s+t}(X.. ,F)$$

natural with respect to X.. *and* F.

Proof. Employing $L^{s\cdot}(\ ) : \mathrm{AbSh}(X_{s.}) \to \mathrm{AbSh}(X..)$ as in the proof of Proposition 2.4, we conclude that $I_{s.} \in \mathrm{AbSh}(X_{s.})$ is injective whenever $I \in \mathrm{AbSh}(X..)$ is injective. We define the complex $Z_{X_{*.}}$ of sheaves on Et(X..) with n-th term equal to $L^{n\cdot}(Z) \in \mathrm{AbSh}(X..)$, so that $Z_{X_{*.}} \to Z$ is a resolution in AbSh(X..). The asserted spectral sequence is that associated to the bicomplex $\mathrm{Hom}_{\mathrm{Absh}(X..)}(Z_{X_{*.}} ,I^{\cdot})$, where $F \to I^{\cdot}$ is an injective resolution in AbSh(X..). ∎

3. COHOMOLOGY VIA HYPERCOVERINGS

The main result of this chapter is Theorem 3.8 which asserts that sheaf cohomology of a simplicial scheme can be computed in a somewhat combinatorial way using hypercoverings. Because the definition of a hypercovering and the required properties of the category of hypercoverings are somewhat formidable at first encounter, we begin this chapter with a discussion of the simpler context of Čech nerves and Čech cohomology. As seen in Corollary 3.9, Čech cohomology is naturally isomorphic to sheaf cohomology in many cases of interest (this is a very special property of the etale site). By considering hypercoverings, we provide combinatorial arguments applicable to any site only notationally more complicated than those of Čech theory (see, for example, Lemma 8.3).

Let $X.$ be a simplicial scheme. An (etale) *covering* $U. \to X.$ of $X.$ is an etale surjective map. The *Čech nerve* of $U. \to X.$ is the bisimplicial scheme $N_{X.}(U.)$ defined by

$$N_{X.}(U..)_{s,t} = (N_{X_s}(U_s))_t$$

the (t+1)-fold fibre product of U_s with itself over X_s (see Example 1.3). For any *abelian presheaf* P (i.e., functor $Et(X.)^0 \to (ab.grps)$), we define $P(N_{X.}(U.))$ to be the bi-cochain complex given in bi-codimension s,t by $P(N_X(U..)_{s,t})$ with differentials obtained in the usual way as an alternating sum of the maps ∂_i^*. We recall that the cohomology of such a bi-complex is the cohomology of the associated total complex

PROPOSITION 3.1. *Let* $X.$ *be a simplicial scheme and* P *an abelian sheaf on* $X.$. *For any* $i \geq 0$, *define the Čech cohomology of* $X.$ *with values in* P *in degree* i *by*

$$\check{H}^i(X.,P) = \operatorname*{colim}_{U.\to X.} H^i(P(N_{X.}(U.)))$$

where the colimit is indexed by coverings $U. \to X.$ *of* $X.$ *So defined,* $\check{H}^*(X,\ )$ *is a* δ*-functor on the category of abelian presheaves on* $X.$.

Proof. Let $U. \to X.$ and $V. \to X.$ be coverings of $X.$ Then $U. \times_{X.} V. \to X.$ defined by $(U. \times_{X.} V.)_n = (U_n \times_{X_n} V_n)$ is also a covering of $X.$ Furthermore, if $f,g : U. \to V.$ are two maps over $X.$, then

$$N_{X.}(f)^* = N_{X.}(g)^* : H^*(P(N_{X.}(V.))) \to H^*(P(N_{X.}(U.))) \quad \text{(see below).}$$

Thus, $\operatorname*{colim}_{U.\to X.}$ can be re-indexed by the opposite category to the *left directed category* (i.e., category with the properties that there is at most one map between any two objects and that for any two objects there exists a third mapping to both) of coverings of $X.$ and equivalence classes of maps, and hence is exact. The fact that $N_{X.}(f)^* = N_{X.}(g)^*$ for $f,g : U. \to V.$ is proved using the fact that two maps $f_n, g_n : U_n \to V_n$ over X_n for $n \geq 0$ have the property that

$$\operatorname{cosk}_0^{X_n}(f_n), \operatorname{cosk}_0^{X_n}(g_n) : \operatorname{cosk}_0^{X_n}(U_n) \to \operatorname{cosk}_0^{X_n}(V_n)$$

are related by a unique simplicial homotopy (since a map $\operatorname{cosk}_0^{X_n}(U_n) \otimes \Delta[1] \to \operatorname{cosk}_0^{X_n}(V_n)$ is equivalent to its restriction to dimension 0). Thus, $N_{X.}(f)$ and $N_{X.}(g)$ are related by a *bi-simplicial homotopy* $N_{X.}(U.) \otimes (\Delta[0] \times \Delta[1]) \to N_{X.}(V.)$, where $(N_{X.}(U.) \otimes (\Delta[0] \times \Delta[1]))_{s,t} = (N_{X_s}(U_s))_t \otimes \Delta[1]_t$, so that $N_{X.}(f)^* = N_{X.}(g)^*$.

By definition, an exact sequence of abelian pre-sheaves $0 \to P_1 \to P_2 \to P_3 \to 0$ on $Et(X.)$ is a sequence with the property that $0 \to P_1(U) \to P_2(U) \to P_3(U) \to 0$ is exact for every $U \to X_n$ in $Et(X.)$. Thus, such a short exact sequence induces a short exact sequence of bi-complexes $0 \to P_1(N_{X.}(U.)) \to P_2(N_{X.}(U.)) \to P_3(N_{X.}(U.)) \to 0$ for any $U. \to X$; this short exact sequence yields a long exact sequence

$$\cdots \to H^i(P_1(N_{X.}(U.))) \to H^i(P_2(N_{X.}(U.))) \to H^i(P_3(N_{X.}(U.))) \to H^{i+1}(P_1(N_{X.}(U.))) \to \cdots .$$

By exactness of $\operatorname{colim}_{U.\to X.}$ as discussed above, we conclude the long exact sequence

$$\cdots \to \check{H}^i(X.,P_1) \to \check{H}^i(X.,P_2) \to \check{H}^i(X.,P_3) \to \check{H}^{i+1}(X.,P_1) \to \cdots .$$

In other words, $\check{H}^*(X.,\ )$ is a δ-functor. ∎

As we will see in the next proposition, there is a spectral sequence analogous to that of Proposition 2.4 which relates Čech cohomology of $X.$ with that of each of the X_n, $n \geq 0$.

PROPOSITION 3.2. *Let* $X.$ *be a simplicial scheme and* P *an abelian presheaf on* $X.$. *Then there exists a first quadrant spectral sequence*

$$E_1^{s,t} = \check{H}^t(X_s, P_s) \Rightarrow \check{H}^{s+t}(X.,P)$$

where P_s *is the restriction of* P *to* $\mathrm{Et}(X_s)$.

Proof. For any covering $U. \to X.$, there is a spectral sequence associated to the bicomplex $P(N_{X.}(U.))$ of the form

$$E_1^{s,t}(U.) = H^t(P_s(N_{X_s}(U_s))) \Rightarrow H^{s+t}(P(N_{X.}(U.))) .$$

Because two maps $U. \rightrightarrows V.$ over $X.$ induce maps $P(N_{X.}(V.)) \rightrightarrows P(N_{X.}(U.))$ which are related by a filtration-preserving homotopy, we conclude that two such maps induce the same map

$$\{E_r^{s,t}(V.)\}_{r\geq 1} \to \{E_r^{s,t}(U.)\}_{r\geq 1} .$$

Therefore, we may take the colimit of the spectral sequence indexed by the left directed category of coverings $U. \to X.$ and equivalence classes of maps to obtain the spectral sequence

$$E_1^{s,t} = \operatorname{colim} H^t(P_s(N_{X_s}(U_s))) \Longrightarrow H^{s+t}(X. ,P) .$$

To conclude the proof of Proposition 3.2, it suffices to verify for any $s,t \geq 0$ that the natural map

$$\operatorname{colim} H^t(P_s(N_{X_s}(U_s))) \to \check{H}^t(X_s,P_s) = \operatorname{colim} H^t(P_s(N_{X_s}(W)))$$

(where the first colimit is indexed by coverings $U. \to X.$ and the second by coverings $W \to X_s$) is an isomorphism. For this, it suffices to observe that if $W \to X_s$ is etale and surjective, then $U_s \to X_s$ factors through $W \to X_s$ where $U. = \Gamma_s^{X.}(W_s) \to X.$ (cf., Prop. 1.5). ∎

The failure of Čech cohomology to equal sheaf cohomology arises from the fact that a family of coverings $U \to X$ can have the property that the U's become "arbitrarily fine" while the $N_X(U)_k$'s do not become arbitrarily fine for some $k \geq 1$. For example, the fact that the U's become acyclic need not imply that the $U \times_X U$'s become acyclic. This problem is circumvented by introducing hypercoverings, the following generalization of Čech nerves.

DEFINITION 3.3. Let X. be a simplicial scheme. An (etale) *hypercovering* $U.. \to X.$ is a bi-simplicial scheme over X. with the property that $U_{s.} \to X_s$ is a hypercovering of X_s for each $s \geq 0$ (i.e., $U_{s,t} \to (\operatorname{cosk}_{t-1}^{X_s} U_{s.})_t$ is etale surjective for all $t \geq 0$, where $\operatorname{cosk}_{-1}^{X_s} U_{s.} = X_s$). The homotopy category of hypercoverings of X. , denoted HR(X.), is the category whose objects are hypercoverings $U.. \to X.$ and whose maps are equivalence classes of maps of hypercoverings of X. (i.e., of bi-simplicial schemes over X.) where the equivalence relation is generated by pairs of maps $U.. \rightrightarrows V..$ related by a simplicial homotopy $U.. \otimes (\Delta[0] \times \Delta[1]) \to V..$ over X. . ∎

We recall that the homotopy category of Čech nerves of coverings $U. \to X.$ is a left directed category. A generalization of left directed category is that of a *left filtering category.* A category C is left filtering if (i) for every pair of objects c, c' in C, there exists c'' in C together with maps $c \leftarrow c'' \rightarrow c'$; and (ii) for every pair of maps $c' \rightrightarrows c$ in C, there exists $c'' \rightarrow c'$ in C such that the composites $c'' \rightarrow c' \rightrightarrows c$ are equal ($c'' \rightarrow c'$ is called a *left equalizer* of $c' \rightrightarrows c$).

In the proof of Proposition 3.2 we used the fact that every covering $W \rightarrow X_s$ is dominated by $U_s \rightarrow X_s$ where $U. = \Gamma_s^{X.}(W)$. We generalize this relationship between the left directed categories of coverings of X. and coverings of X_s as follows. A functor $F : C \rightarrow D$ is said to be *left final* provided that (i) for every object d of D, there exists an object c in C and a map $F(c) \rightarrow d$ in D; and (ii) for every object c in C and every pair of maps $F(c) \rightrightarrows d$ in D, there exists $\gamma : c' \rightarrow c$ in C such that $F(\gamma)$ is a left equalizer of the given maps $F(c) \rightrightarrows d$.

The usefulness of these generalizations is that if C is left filtering, then the colimit indexed by the opposite category of C, $\operatorname{colim}_{C^0}$, is an exact functor on the category of abelian group valued functors on C^0; moreover, if C and D are left filtering, $F : C \rightarrow D$ is left final, and $P : D^0 \rightarrow Ab$ is any functor, then the natural map $\underset{\overrightarrow{C^0}}{\operatorname{colim}}\, P \circ F \rightarrow \underset{\overrightarrow{D^0}}{\operatorname{colim}}\, P$ is an isomorphism (cf., [8], A.1.8).

Proposition 3.4 reveals the relevance of this discussion to the category HR(X.) of Definition 3.3.

PROPOSITION 3.4. *Let* X. *be a simplicial scheme. Then* HR(X.) *is left filtering and the restriction map* $HR(X.) \rightarrow HR(X_n)$ *is left final for each* $n \geq 0$.

Proof. If $U.. \rightarrow X.$ and $V.. \rightarrow X.$ are hypercoverings, then $U.. \times_{X.} V.. \rightarrow X.$ (defined by $(U.. \times_{X.} V..)_{s,t} = U_{s,t} \times_{X_s} V_{s,t}$) is a hypercovering mapping to $U.. \rightarrow X.$ and $V.. \rightarrow X.$. If $U.. \rightrightarrows V..$ are two maps of hypercoverings of X. , we use the construction of ([7], V.7.3.7) to construct the left equalizer. Namely, for each $s \geq 0$, there is a naturally defined simplicial scheme $W_{s.} = \mathcal{H}om(X_s \otimes \Delta[1], V_{s.})$ over X_s with the property that

$$\mathrm{Hom}(Z., \mathcal{H}om(X_s \otimes \Delta[1], V_{s.})) = \mathrm{Hom}(Z. \otimes \Delta[1], V_{s.})$$

for any simplicial scheme $Z.$ over X_s, where Hom(,) denotes maps of simplicial schemes over X_s. Furthermore, the natural map

$$W_{s.} \to V_{s.} \times_{X_s} V_{s.}$$

is *special* over X_s (where $Z. \to T.$ is special over Y if Z_n maps surjectively onto the fibre product of $(\mathrm{cosk}^Y_{n-1} Z.)_n \to (\mathrm{cosk}^Y_{n-1} T.) \leftarrow T_n$ for all $n \geq 0$), as seen for example in [29], 1.3. Consequently, the fibre product $U'_{..}$ of $W.. \to V.. \times_{X.} V.. \leftarrow U..$ is a hypercovering of $X.$ mapping to $U..$. Moreover, by construction, the two compositions $U'_{..} \to U.. \rightrightarrows V..$ are related by a simplicial homotopy because they factor through $W.. \rightrightarrows V..$. Thus, $U'_{..} \to U..$ is a left equalizer of $U.. \rightrightarrows V..$.

To prove that $HR(X.) \to HR(X_n)$ is left final, we use Proposition 1.5 to conclude that $\Gamma^{X.}_n(\ )$ sending $W.$ to $\Gamma^{X.}_n(W.)$ (defined by $(\Gamma^{X.}_n(W.))_{s,t} = (\Gamma^{X.}_n(W_t))_s$) is right adjoint to the restriction functor from bisimplicial schemes over $X.$ to simplicial schemes over X_n. Moreover, if $W.$ is a hypercovering of X_n, then $\Gamma^{X.}_n(W.)$ is readily checked to be a hypercovering of $X.$ (because $(\Gamma^{X.}_n(W.))_{s.}$ is the fiber product over X_s of pull-backs of $W. \to X_n$). Thus, the left finality of $HR(X.) \to HR(X_n)$ follows from the following lemma (whose proof we leave as an exercise). ∎

LEMMA 3.5. *Let* $F: C \to D$ *be a functor inducing a functor* $\mathrm{Ho}F: \mathrm{Ho}C \to \mathrm{Ho}D$ *where maps in* $\mathrm{Ho}C$ *and* $\mathrm{Ho}D$ *are equivalence classes of maps of* C *and* D *respectively. If* $\mathrm{Ho}C$ *and* $\mathrm{Ho}D$ *are left filtering and if* F *admits a right adjoint, then* $\mathrm{Ho}F$ *is left final.* ∎

The next proposition indicates that a hypercovering of a scheme should be viewed as a resolution of that scheme.

PROPOSITION 3.6. *Let* X *be a scheme and* $U. \to X$ *a hypercovering. Let* Z_{U_*} *be the complex of abelian sheaves in* AbSh(X) *determined by* $U. \to X$ *whose* m*-th term is the abelian sheaf represented by* $U_m \to X$ *in*

Et(X). *Then the natural augmentation map* $Z_{U_*} \to Z$ *induced by* $U_0 \to X$ *expresses* Z_{U_*} *as a resolution in* AbSh(X) *of* Z.

Proof. To prove that $Z_{U_*} \to Z \to 0$ is exact in AbSh(X), it suffices to show that $x^*(Z_{U_*} \to Z \to 0)$ in AbSh(Spec k) is exact for every geometric point $x : \operatorname{Spec} k \to X$ (by Proposition 2.2). Fix a geometric point $x : \operatorname{Spec} k \to X$ and let $W. = U. \times_X \operatorname{Spec} k \to \operatorname{Spec} k$ denote the pull-back of $U. \to X$ by x. Then $x^*(Z_{U_*} \to Z \to 0)$ equals $Z_{W_*} \to Z \to 0$ in AbSh(Spec k). Because Et(Spec k) has $\operatorname{id} : \operatorname{Spec} k \to \operatorname{Spec} k$ as an initial object, the global section functor AbSh(Spec k) $\to$ Ab is an equivalence. Yet, $Z_{W_*}(\operatorname{Spec} k)$ is the free abelian chain complex $C_*(W.)$ on the simplicial set W. (ignoring the scheme structure). Thus, it suffices to prove that $C_*(W.) \to Z \to 0$ is exact. This follows from the observation that W. is the contractible Kan complex because it satisfies the hypercovering condition that $W_t \to (\operatorname{cosk}_{t-1} W.)_t$ be surjective (as a map of sets) for all $t \geq 0$. ■

Using Proposition 3.6, we prove that a simplicial scheme and its hypercoverings have the same cohomology.

PROPOSITION 3.7. *Let* X. *be a simplicial scheme and* $U.. \to X.$ *be a hypercovering. Then* $\Delta U.. \to X.$ *induces an isomorphism of* δ*-functors on* AbSh(X.)

$$H^*(X. , \ ) \xrightarrow{\sim} H^*(U.. , \ ) .$$

Proof. If $F \in$ AbSh(X.), then $H^0(X. ,F) = \operatorname{Ker}(F(X_0) \rightrightarrows F(X_1))$, whereas $H^0(U.. ,F) = \operatorname{Ker}(F(U_{0,0}) \rightrightarrows F(U_{1,1})) = \operatorname{Ker}(d_0^* - d_1^* : F(U_{0,0}) \to F(U_{1,1}))$. Because $U_{n.} \to X_n$ is a hypercovering for each $n \geq 0$, the sheaf axiom implies that $\operatorname{Ker}(F(U_{n,0}) \rightrightarrows F(U_{n,1})) = \operatorname{Ker}(F(U_{n,0}) \rightrightarrows F(U_{n,0} \times_{X_n} U_{n,0}))$ equals $F(X_n)$. Consequently,

$$H^0(U.. ,F) = \operatorname{Ker}(F(U_{0,0}) \rightrightarrows F(U_{1,1})) \xrightarrow{\sim} \operatorname{Ker}(F(X_0) \rightrightarrows F(X_1)) = H^0(X. ,F).$$

Since the restriction functor $\mathrm{AbSh}(X.) \to \mathrm{AbSh}(U..)$ is exact, $H^*(U.., \;)$ is a δ-functor on $\mathrm{AbSh}(X.)$. Thus, it suffices to prove that $H^*(U.., I) = H^*(X., I)$ for I injective in $\mathrm{AbSh}(X.)$. Because $U_{s,t} \to X_s$ is etale, the restriction functor has an exact left adjoint $\gamma: \mathrm{Ab}(U_{s,t}) \to \mathrm{Ab}(X_s)$ defined by sending a sheaf G on $U_{s,t}$ to the sheaf associated to the presheaf $(V \to X_s) \mapsto \oplus G(V \to U_{s,t})$, where the sum is indexed by the set of maps $V \to U_{s,t}$ over X_s. Consequently, the restriction to $U_{s,t}$ of an injective on X_s is injective. Therefore, $H^*(U.., I) = H^*(\Delta U.., I) = H^*(I(\Delta U..))$ by Propositions 2.4 and 2.5. To compute $H^*(I(\Delta U..)) = H^*(I(U..))$, we employ the spectral sequence

$$E_1^{s,t} = H^t(I_s(U_{s.})) \Rightarrow H^{s+t}(I(U..)) .$$

Because I_s is injective in $\mathrm{AbSh}(X_s)$ and because $I_s(U_{s.}) = \mathrm{Hom}_{\mathrm{AbSh}(X_s)}(Z_{U_{s*}}, I^s)$, Proposition 3.6 implies that $H^t(I_s(U.)) = 0$ for $t > 0$ and $H^0(I_s(U_{s.})) = I_s(X_s)$ for each $s \geq 0$. Consequently, $H^*(I(U..)) = H^*(I(X.))$ which equals $H^*(X., I)$ by the remark following Proposition 2.4. ■

We now prove that sheaf cohomology can be computed using hypercoverings. For schemes, this theorem was first proved by J.-L. Verdier in [7], V.7.4.1.

THEOREM 3.8. *Let* $X.$ *be a simplicial scheme. Then there is a natural isomorphism of* δ*-functors on* $\mathrm{AbSh}(X.)$

$$H^*(X., \;) \xrightarrow{\sim} \operatorname{colim} H^*((U..))$$

where the colimit is indexed by $U..$ *in* $HR(X.)$.

Proof. As seen in the proof of Proposition 3.7, for any $F \in \mathrm{AbSh}(X.)$ and any $V..$ in $HR(X.)$, $H^0(X., F) \xrightarrow{\sim} H^0(V.., F) = H^0(F(V..)) \simeq \operatorname{colim} H^0(F(U..))$; moreover, for any injective $I \in \mathrm{AbSh}(X.)$ and any $V..$ in $HR(X.)$, $H^*(X., I) \xrightarrow{\sim} H^*(V.., I) \xrightarrow{\sim} H^*(I(V..)) \xrightarrow{\sim} \operatorname{colim} H^*(I(U..))$. Thus, it suffices to prove that a short exact sequence $0 \to F_1 \to F_2 \to F_3 \to 0$ in AbSh(X.) determines a long exact sequence

$$\cdots \to \operatorname{colim} H^*(F_1(U..)) \to \operatorname{colim} H^*(F_2(U..)) \to \operatorname{colim} H^*(F_3(U..)) \to \cdots .$$

Let $G_1 = \operatorname{coker}(F_1 \to F_2)$ as presheaves on Et(X.), and let $G_3 = \operatorname{coker}(G_1 \to F_3)$, so that $0 \to G_1 \to F_3 \to G_3 \to 0$ is a short exact sequence of presheaves on Et(X.). For each U.., $0 \to F_1(U..) \to F_2(U..) \to G_1(U..) \to 0$, $0 \to G_1(U..) \to F_3(U..) \to G_3(U..) \to 0$ are exact, thus yielding long exact sequences in cohomology. Because HR(X.) is left filtering and because these long exact sequences are functorial on HR(X.), we conclude that it suffices to prove that $\operatorname{colim} H^*(G_3(U..)) = 0$.

Let $a \in G_3(U_{n,n})$ represent a given cohomology class α in $H^n(G_3(U..))$, some $n \geq 0$, and U.. in HR(X.). Because the sheaf associated to the presheaf G_3 is the zero sheaf, we may choose an etale surjective map $W \to U_{n,n}$ such that a restricts to 0 in $G_3(W)$. We claim that $\Gamma_n^{U_{n.}}(W) = V. \to U_{n.}$ is a special map over X_n (see Proposition 1.7). Namely, the appropriate surjectivity in dimension $t \geq 0$ is equivalent to the lifting of geometric points, which is equivalent to lifting maps from $\operatorname{Spec} k \otimes \Delta[t]$. This can be readily checked using the definition of $\Gamma_n^{U.}(\ )$ and the adjointness of $\operatorname{cosk}_{t-1}(\ )$ and $\operatorname{sk}_{t-1}(\ )$. Consequently, $U'.. = \Gamma_n^{X.}(V.) \to \Gamma_n^{X.}(U_{n.})$ is a special map over X., so that the fibre product $U''..$ of $U'.. \to \Gamma_n^{X.}(U_{n.}) \leftarrow U..$ is a hypercovering of X. such that $U''_{n,n} \to U_{n,n}$ factors through $W \to U_{n,n}$. Therefore, a restricts to 0 in $G_3(U''_{n,n})$, so that α goes to 0 in $\operatorname{colim} H^n(G_3(U..))$. ∎

As a corollary of Theorem 3.8, we generalize M. Artin's theorem that sheaf cohomology equals Čech cohomology in many cases of interest [6]. This result is special to the etale topology, failing even for sheaves on the Zariski topology.

COROLLARY 3.9. *Let* X. *be a simplicial scheme such that* X_n *is quasi-projective over a noetherian ring for each* $n \geq 0$. *Then there is a natural isomorphism of* δ*-functors on* AbSh(X.)

$$\check{H}^*(X. , \ ) \xrightarrow{\simeq} H^*(X. , \ ) .$$

Proof. The inclusion of the homotopy category of Čech nerves of coverings $V. \to X.$ into the homotopy category HR(X.) of hypercoverings $U.. \to X.$ induces a natural transformation $\check{H}^*(X., \ ) = \operatorname{colim} H^*((N_{X.}(V.))) \to \operatorname{colim} H^*((U..)) = H^*(X., \ )$ of δ-functors on AbSh(X.). For any $F \in$ AbSh(X.), this map is induced by a map of filtered complexes, and thus may be identified with the map on abutments of a map of spectral sequence

$$\begin{array}{ccc} E_1^{s,t} = \check{H}^t(X_s,F_s) & \Rightarrow & \check{H}^{s+t}(X. ,F) \\ \downarrow & & \\ {}'E_1^{s,t} = H^t(X_s,F_s) & \Rightarrow & H^{s+t}(X. ,F) \end{array}$$

where the first spectral sequence is that of Proposition 3.2 and the $'E_1$-term is identified using the isomorphism $H^*(X_s,F_s) \xrightarrow{\simeq} \operatorname{colim} H^*(F_s(U_{s.}))$ provided by the left finality of $HR(X.) \to HR(X_s)$. The corollary now follows by applying Artin's theorem to conclude that this map of spectral sequences is an isomorphism. ∎

For the sake of completeness, we determine the effect of $\operatorname{colim} H^*((U..))$ on abelian presheaves with the aid of Theorem 3.8.

COROLLARY 3.10. *Let* X. *be a simplicial scheme, and let* $P : Et(X.)^0 \to$ (ab. grps.) *be an abelian presheaf with the property that* $P(\coprod_{i \in I} U_i) = \prod_{i \in I} P(U_i)$ *for any* $\coprod_{i \in I} U_i \to X_n$ *in* Et(X.). *Let* $P^{\#} : Et(X.)^0 \to$ (ab. grps.) *denote the sheaf associated to* P (*cf.* [59], II.2.11). *Then there is a natural isomorphism*

$$H^*(X.\,,P^\#) \longrightarrow \operatorname{colim} H^*(P(U..))$$

where the colimit is indexed by $U.. \to X.$ *in* $HR(X.)$.

Proof. By Theorem 3.8, it suffices to prove that the natural map $P \to P^\#$ determines an isomorphism of chain complexes

$$\operatorname{colim} P(U..) \to \operatorname{colim} P^\#(U..)\ .$$

Because P and $P^\#$ have isomorphic stalks at every geometric point and because P "commutes" with disjoint unions, we conclude the following: If $\alpha \in P(U_{s,t})$ goes to $0 \in P^\#(U_{s,t})$, then there exists an etale surjective map $U' \to U_{s,t}$ such that α restricts to $0 \in P(U')$; and if $\beta \in P^\#(U_{s,t})$, then there exists an etale surjective map $U'' \to U_{s,t}$ such that the restriction of β in $P^\#(U'')$ is in the image of $P(U'')$. As argued in the proof of Theorem 3.8, given any etale surjective map $W \to U_{s,t}$ there exists a map of hypercoverings $U'.. \to U..$ such that $U'_{s,t} \to U_{s,t}$ factors through $W \to U_{s,t}$. Consequently, $\operatorname{colim} P(U_{s,t}) \to \operatorname{colim} P^\#(U_{s,t})$ is an isomorphism for any $s,t \geq 0$ as required. ∎

4. ETALE TOPOLOGICAL TYPE

As we observed in the last chapter, the sheaf cohomology of a simplicial scheme X. is determined by its hypercoverings. The etale topological type of X., $(X.)_{et}$, is essentially the inverse system of simplicial sets given by applying the connected component functor to the hypercoverings. As we see in Chapter 5, the sheaf cohomology of X. with locally constant coefficients can be computed from the homotopy type of $(X.)_{et}$.

Our use of general hypercoverings rather than only Čech nerves does complicate our construction, although this complication is primarily one of notation. The reader is urged to consider only Čech nerves when initially considering the etale topological type: the inverse system of simplicial sets associated to Čech nerves (the Čech topological type) is shown in Proposition 8.2 to be weakly equivalent to $(X.)_{et}$ in most cases of interest. On the other hand, there are several important advantages of our definition of the etale topological type using general hypercoverings. Namely, $(X.)_{et}$ has the "correct" weak homotopy type for all locally noetherian simplicial schemes and $(\ )_{et}$ is indeed a refinement of Artin-Mazur's etale homotopy type (Proposition 4.5). Perhaps most important, our construction applies to other sites (e.g., the Zariski site) in which Čech cohomology differs from derived functor cohomology. This construction of the etale topological type is based on that of the Čech topological type first appearing in [34] and [51]. The essential idéa, suggested by work of S. Lubkin [54], is to "rigidify" coverings by providing each connected component with a distinguished geometric point.

Because the etale topological type is not a single simplicial set (or space), etale homotopy theory is somewhat off-putting at first glance. As defined in Definition 4.4, $(X.)_{et}$ is a pro-simplicial set (a functor from a left filtering category to simplicial sets). Although pro-objects were first systematically employed in etale homotopy (cf. [8], Appendix), the special case of pro-groups has been widely used in galois cohomology [66]. At first acquaintance with such pro-objects, one is tempted to take an inverse limit. Not only does one lose structure once the inverse limit is applied, but also one obtains incorrect invariants: the cohomology of a pro-simplicial set is a kind of continuous cohomology of its inverse limit.

The knowledgeable reader will recognize the close analogy of our construction of $(X.)_{et}$ to shape theory in topology [28]. In fact, if one considers a "nice" topological space T (e.g., a finite C.W. complex) and a "hypercovering" of T with each component a contractible open subset of T, then the simplicial set of connected components of this hypercovering has the "same" homotopy type as X ([8], 12.1).

We begin with the following proposition which provides the rigidity of our construction. The topological analogue of Proposition 4.1 is the fact that a map of connected covering spaces $X' \to X''$ of a space X is determined by its value on a single point.

PROPOSITION 4.1. *Let* X *be a scheme,* $U \to X$ *etale with* U *connected, and* $V \to X$ *etale and separated. Then two maps* $f,g : U \to V$ *over* X *are equal if* $f \circ u = g \circ u$ *for some geometric point* $u : \operatorname{Spec} k \to U$.

Proof. We interpret f and g as sections of the projection $U \underset{X}{\times} V \to U$. Such sections are open ([59], I.3.12) and closed (because $V \to X$ is separated). Because U is connected, we thus may identify a section with a choice of connected components of $U \underset{X}{\times} V$ isomorphic to U. Because u and $f \circ u$ (respectively, $g \circ v$) determine a geometric point of the connected component of $U \underset{X}{\times} V$ corresponding to f (respectively, g), f equals g whenever $f \circ u$ equals $g \circ v$. ■

The reader should beware that Proposition 4.1 is false if one merely assumes that f and g agree on a scheme-theoretic point. For example, take $X = \operatorname{Spec} k$ and $U = V = \operatorname{Spec} K$ with K/k a finite galois extension; then each distinct automorphism of K over k determines a distinct map from U to V, but these distinct maps of course agree on the unique scheme theoretic point of U.

We next introduce the definition of a rigid (etale) covering of a scheme X. Because we shall employ the uniqueness property of Proposition 4.1, we index the connected components by the geometric points of X and provide each connected component with a distinguished geometric point. A first problem arises in that an etale open U of X may not be a disjoint union of its connected components. A second problem is the awkwardness that the collection of all geometric points of X is not a set.

We recall that a scheme X is said to be *locally noetherian* provided that it is a union of affine noetherian schemes (i.e., spectra of noetherian rings). Because the Zariski topological space of a locally noetherian scheme has the property that every point has a system of neighborhoods in which each descending chain of closed subsets is finite, we readily conclude that each scheme-theoretic point of a locally noetherian scheme has such a neighborhood which is also irreducible. This easily implies that the connected components of X are open as well as closed ([47], I. 6.1.9).

To avoid set theoretic problems, we choose for each characteristic $p \geq 0$ an algebraically closed field Ω_p which is sufficiently large to contain subfields isomorphic to the residue fields of characteristic p of every scheme we consider. (At this point, set theorists would suggest we fix a "universe" and consider only those simplicial schemes which in each dimension lie in this universe.) Once such a choice is made, we define the *set of geometric points* of X, denoted $\overline{X}$, to consist of all geometric points $x : \operatorname{Spec} \Omega \to X$ where $\Omega = \Omega_p$ with p the residue characteristic of the image scheme theoretic point of X.

With these preliminaries, we introduce rigid coverings.

DEFINITION 4.2. A *rigid covering* $\alpha : U \to X$ of a locally noetherian scheme X is a disjoint union of pointed etale, separated maps

$$\alpha_x : U_x, u_x \to X, x, \qquad \forall x \in \overline{X}$$

where each U_x is connected and $\alpha_x \circ u_x = x : \mathrm{Spec}\ \Omega \to X$. A *map of rigid coverings* over a map $f : X \to Y$ of schemes, $\phi : (\alpha : U \to X) \to (\beta : V \to Y)$ is a map $\phi : U \to V$ over f such that $\phi \circ u_x = v_{f(x)}$ for all $x \in \overline{X}$. If $U \to X$ and $V \to X$ are rigid coverings of X and Y over a third scheme S, then the *rigid product* $U \overset{R}{\underset{S}{\times}} V \to X \underset{S}{\times} Y$ is defined to be the open and closed immersion of $U \underset{S}{\times} V \to X \underset{S}{\times} Y$ given as the disjoint union indexed by geometric points $x \times y$ of $X \underset{S}{\times} Y$ of

$$\alpha_x \times \beta_y : (U_x \underset{S}{\times} V_y)_0, \ u_x \times v_y \to X \underset{S}{\times} Y, \ x \times y$$

where $(U_x \underset{S}{\times} V_y)_0$ is the connected component of $U_x \underset{S}{\times} V_y$ containing $u_x \times v_y$. ■

Of course, Proposition 4.1 implies that there is at most one map between rigid coverings. This implies that $RC(X.)$, the *category of rigid coverings* of a simplicial scheme $X.$, is a left directed category. We proceed to define rigid hypercoverings in such a way that they too determine a left directed category.

PROPOSITION 4.3. *Let* $X.$ *be a locally noetherian simplicial scheme. A rigid hypercovering* $U.. \to X.$ *is a hypercovering with the property that*

$$U_{s,t} \to (\mathrm{cosk}_{t-1}^{X_s} U_{s.})_t$$

is a rigid covering for each $s,t \geq 0$ *such that any map* $\alpha : \Delta[s'] \to \Delta[s]$

induces a map of rigid coverings over $\alpha : (\mathrm{cosk}_{t-1}^{X_s} U_{s.})_t \to (\mathrm{cosk}_{t-1}^{X_{s'}} U_{s'.})_t$ *for each* $t \geq 0$. *A map of rigid hypercoverings over a map* $f : X. \to Y.$ *is a map of bisimplicial schemes* $\phi : U.. \to V..$ *over* f *such that* $\phi_{s,t} : U_{s,t} \to V_{s,t}$ *is a map of rigid coverings over* $\mathrm{cosk}_{t-1}\phi : (\mathrm{cosk}_{t-1} U_{s.}^{X_s})_t \to (\mathrm{cosk}_{t-1}^{Y_s} V_{s.})_t$ *for each* $s,t \geq 0$. *The category of rigid hypercoverings of* $X.$, $\mathrm{HRR}(X.)$, *is a left directed category.*

Proof. Because any simplicial map $\phi_{s.} : U_{s.} \to V_{s.}$ has the property that the (t-1)-truncation of $\phi_{s.}$ determines $\mathrm{cosk}_{t-1}\phi_{s.}$, we conclude by Proposition 4.1 that there is at most one map between any two rigid hypercoverings over a given map of simplicial schemes. To prove $\mathrm{HRR}(X.)$ is left directed, it thus suffices to observe that if $U.. \to X.$ and $V.. \to X.$ are rigid hypercoverings then their *rigid product* $U.. \overset{R}{\underset{X.}{\times}} V.. \to X.$ is a rigid hypercovering mapping to both, where $(U.. \overset{R}{\underset{X.}{\times}} V..)_{s,t} \to (\mathrm{cosk}_{t-1}(U_{s.} \overset{R}{\underset{X_s}{\times}} V_{s.}))_t$ is defined to be the restriction to $(\mathrm{cosk}_{t-1}(U_{s.} \overset{R}{\underset{X_s}{\times}} V_{s.}))_t \subset (\mathrm{cosk}_{t-1} U_{s.})_t \underset{X_s}{\times} (\mathrm{cosk}_{t-1} V_{s.})_t$ of the rigid product over X_s of $U_{s,t} \to (\mathrm{cosk}_{t-1} U_{s.})_t$ and $V_{s,t} \to (\mathrm{cosk}_{t-1} V_{s.})_t$. ∎

We remind the reader that a *pro-object* of a category C is a functor from some small left filtering category to C. We shall often denote a pro-object $F : I \to C$ by $\{F_i ; i \in I\}$ or simply $\{F_i\}$, and we shall usually ignore the set theoretic requirement that I be a small category. A *map of pro-objects* from $F : I \to C$ to $G : J \to C$ is an element in the set $\varprojlim_J \operatorname{colim}_I \mathrm{Hom}_C(F(i), G(j))$; in other words, a map from F to G is a compatible collection of elements (indexed by J) in $\operatorname{colim}_I \mathrm{Hom}_C(F(i), G(j))$, each of which is determined by a map

$F(i) \to G(j)$ for some i (depending on $j \in J$). A *strict map of pro-objects* from $F : I \to C$ to $G : J \to C$ is a pair consisting of a functor $\alpha : J \to I$ and a natural transformation $\phi : F \circ \alpha \to G$. Of course, a strict map of pro-objects determines a map of pro-objects.

If $F : I \to C$ is a pro-object and if $\alpha : J \to I$ is a functor between left filtering categories, then $(\alpha : J \to I,\ \mathrm{id} : F \circ \alpha \to F \circ \alpha)$ is a strict map from F to $F \circ \alpha$. If the functor $\alpha : J \to I$ is left final, then we readily conclude the existence of a (not necessarily strict) map of pro-objects $F \circ \alpha \to F$ inverse to this canonical map. In other words, the strict map $F \to F \circ \alpha$ is an isomorphism of pro-objects, even though it may not have a strict inverse.

The *Čech topological type* of $X.$ is defined to be the pro-simplicial set $\Delta(X.)_{ret} = \pi \Delta N_{X.} : RC(X.) \to (\text{s. sets})$ (notation is that of [34]) sending a rigid covering $U.. \to X.$ to the simplicial set $\pi(\Delta(N_{X.}(U.)))$ given in dimension t as the set of connected components of the t-fold fibre product of U_t over X_t. This definition suggests our definition of the etale topological type.

DEFINITION 4.4. Let $X.$ be a locally noetherian simplicial scheme. The *etale topological type* of $X.$ is defined to be the following pro-simplicial set

$$(X.)_{et} = \pi \circ \Delta : HRR(X.) \to (\text{s. sets})$$

sending a hypercovering $U..$ of $X.$ to the simplicial set $\pi(\Delta U..)$ of connected components of the diagonal of $U..$ (so that $\pi(\Delta U..)_n$ is the set of connected components of $U_{n,n}$). If $f : X. \to Y.$ is a map of locally noetherian simplicial schemes, then the *strict etale topological type* of f is the strict map

$$f_{et} : (X.)_{et} \to (Y.)_{et}$$

given by the functor $f^* : HRR(Y.) \to HRR(X.)$ and the natural transformation $(X.)_{et} \circ f^* \to (Y.)_{et}$ induced by the natural maps $f^*(V..) \to V..$ for $V.. \to Y.$ in $HRR(Y.)$. The etale topological type functor

$$(\ )_{et} : (\text{loc. noeth. s. schemes}) \to (\text{pro-s. sets})$$

sends a locally noetherian simplicial scheme $X.$ to $(X.)_{et}$ and a map $f: X. \to Y.$ to the map of pro-simplicial sets determined by f_{et}. ■

In the above definition, the functor $f^*: \mathrm{HRR}(Y.) \to \mathrm{HRR}(X.)$, the *rigid pull-back* functor, is defined in terms of the pull-backs of rigid coverings. If $V \to Y$ is a rigid covering and $f: X \to Y$ is a map, then $f^*(V \to Y) = U \to X$ is the disjoint union of pointed maps $(V_{f(x)} \underset{Y}{\times} X)_0,\ f(x) \times x \to X, x$ where $(V_{f(x)} \underset{Y}{\times} X)_0$ is the connected component of $V_{f(x)} \underset{Y}{\times} X$ containing the geometric point $f(x) \times x$. If $V.. \to Y.$ is a hypercovering of $Y.$, then $f^*(V.. \to Y.) = U.. \to X.$ is defined for any $s,t \geq 0$ by

$$U_{s,t} \to (\operatorname{cosk}_{t-1}^{X_s} U_{s.})_t = f^*(V_{s,t} \to (\operatorname{cosk}_{t-1}^{Y_s} V_{s.})) .$$

If $X.$ is a simplicial scheme provided with a chosen geometric point $x: \operatorname{Spec} \Omega \to X_0$, then $X.$ (or, more precisely $(X., x)$) is said to be a *pointed simplicial scheme*. Clearly, if $X.$ is a pointed locally neotherian simplicial scheme, then $(X.)_{et}$ is naturally a pro-object in the category (s. sets$_*$) of pointed simplicial sets. Moreover, if $f: X. \to Y.$ is a pointed map of pointed locally noetherian simplicial schemes, then f_{et} is naturally a strict map of pro-(s. sets$_*$).

Originally, the *etale homotopy type* of a scheme X was defined to be $X_{ht} = \pi \circ \Delta: \mathrm{HR}(X) \to \mathcal{H}$, where $\mathcal{H}$ is the homotopy category of simplicial sets defined by inverting weak equivalences ([8], 9.6).

PROPOSITION 4.5. *Let* X *be a locally noetherian scheme, let* $(X \otimes \Delta[0])_{et} \in$ pro-(s. sets) *be defined as in Definition 4.4, and let* $X_{ht} \in$ pro-$\mathcal{H}$ *be defined as above. If* $(X \otimes \Delta[0])_{et}$ *is viewed in* pro-$\mathcal{H}$ *by applying the forgetful functor* (s. sets) $\to \mathcal{H}$, *then* $(X \otimes \Delta[0])_{et}$ *is isomorphic to* X_{ht} *in* pro-$\mathcal{H}$.

Proof. We readily verify that a rigid hypercovering $U.. \to X \otimes \Delta[0]$ is of the form $(U. \to X) \otimes \Delta[0]$, where $U. \to X$ is a rigid hypercovering of X

(i.e., $U_n \to (\cosk_{n-1}U.)_n$ is a rigid covering for $n \geq 0$). Thus, it suffices to verify that $HRR(X) \to HR(X)$ is left final, where $HRR(X)$ is the left directed category of rigid hypercoverings of X.

If $g: U \to V$ is a surjective map, then a "choice of right inverse to g_*" is a function $g_*^{-1}: \overline{V} \to \overline{U}$ right inverse to $g_*: \overline{U} \to \overline{V}$ sending $u: \operatorname{Spec} \Omega \to U$ to $g_*(u) = u \circ g: \operatorname{Spec} \Omega \to V$. If $g: U \to V$ is etale and surjective, and if g_*^{-1} is a choice of right inverse to g_*, we let

$$R(g,g_*^{-1}) \to V = \coprod_{v \in \overline{V}} (U_v, g_*^{-1}(v) \to V,v)$$

be a rigid covering of V, where U_v is a connected, separated Zariski open of U containing $g_*^{-1}(v)$. Clearly, $R(g,g_*^{-1}) \to V$ factors naturally through $g: U \to V$.

Using this construction, we exhibit for $U. \to X$ in $HR(X)$, $R(U.) \to X$ in $HRR(X)$ naturally mapping to $U. \to X$. Define $R(U.)_0 \to X$ to be $R(g_0,g_{0*}^{-1}) \to X$, where $g_0: U_0 \to X$ is the augmentation map and g_{0*}^{-1} is a choice of right inverse to g_{0*}. Inductively, define $R(U.)_n \to (\cosk_{n-1}R(U.))_n$ to be $R(g_n,g_{n*}^{-1}) \to (\cosk_{n-1}R(U.))_n$ where g_n is the projection of the fibre product of $(\cosk_{n-1}R(U.))_n \to (\cosk_{n-1}U.)_n \leftarrow U_n$ onto $(\cosk_{n-1}R(U.))_n$ and g_{n*}^{-1} is induced by the degeneracy maps on $(\operatorname{sk}_{n-1}R(U.))_n \subset (\cosk_{n-1}R(U.))_n$.

To complete the proof of left finality, let $U. \rightrightarrows V.$ be two maps of simplicial schemes with $U.$ in $HRR(X)$ and $V.$ in $HR(X)$, and let $h: H. \to U.$ be the left equalizer constructed in the proof of Proposition 3.4. Using the fact that $H. \to U.$ is special, we construct a map $f: W. \to H.$ in $HR(X.)$ such that $h \circ f: W. \to U.$ is in $HRR(X)$.

We define $f_0: W_0 \to H_0$ to be $R(\varepsilon,\varepsilon_*^{-1}) \to H_0$, where $\varepsilon: H_0 \to X$ is the augmentation map and ε_*^{-1} satisfies $h_* \circ \varepsilon_*^{-1}(x) = u_x$ for all $x \in \overline{X}$ (with u_x the distinguished geometric point above x of the rigid covering $U_0 \to X$). With ε_*^{-1} so chosen, $h_0 \circ f_0: W_0 \to U_0$ is a map of rigid coverings of X. Inductively, we let $g_n': K_n \to L_n$ denote the map induced by h from the fibre product K_n of $H_n \to (\operatorname{cosk}_{n-1}H.)_n \leftarrow (\operatorname{cosk}_{n-1}W.)_n$ to the fibre product L_n of $U_n \to (\operatorname{cosk}_{n-1}U.)_n \leftarrow (\operatorname{cosk}_{n-1}W.)_n$. The rigidity of U. provides a function $\overline{(\operatorname{cosk}_{n-1}W.)_n} \to \overline{L}_n$; because h is special, g_n' is surjective so that a choice of right inverse $g_{n*}'^{-1}$ to g_{n*}' determines a choice of right inverse g_{n*}^{-1} to g_{n*}, where $g_n = pr_2 \circ g_n' : K_n \to (\operatorname{cosk}_{n-1}W.)_n$. We define $W_n \to (\operatorname{cosk}_{n-1}W.)_n$ to be $R(g_n, g_{n*}^{-1})$ and $f_n : W_n \to H_n$ to be induced by $K_n \to H_n$; so determined $W_n \to U_n$ is a map of rigid coverings. ■

Although the constructions of the proof of Proposition 4.5 are not sufficiently natural to prove that HRR(X.) → HR(X.) is left final, we can conclude that sheaf cohomology of X. can be computed using rigid hypercoverings (thanks to Theorem 3.8).

COROLLARY 4.6. *Let X. be a locally noetherian simplicial scheme. Then the forgetful functor induces an isomorphism of δ-functors on* AbSh(X.)

$$\operatorname*{colim}_{U..\in HRR(X.)} H^*(\ (U..)) \to \operatorname*{colim}_{V..\in HR(X.)} H^*(\ (V..)) .$$

Proof. For any F ∈ AbSh(X.), the forgetful functor induces a map of spectral sequences

$$\begin{array}{ccc} E_1^{p,q} = \operatorname*{colim}_{U..} H^q(F(U_{p.})) & \Longrightarrow & \operatorname*{colim}_{U..} H^{p+q}(F(U..)) \\ & \downarrow & \\ 'E_1^{p,q} = \operatorname*{colim}_{V..} H^q(F(V_{p.})) & \Longrightarrow & \operatorname*{colim}_{V..} H^{p+q}(F(V..)) \end{array}$$

We show that this map is an isomorphism. By Propositions 3.4 and 4.5, it suffices to prove that the restriction map $HRR(X.) \to HRR(X_p)$ is left final for each $p \geq 0$. Imitating the proof of Proposition 3.4, we define $R\Gamma_p^{X.}(W.) \to X.$ in HRR(X.) associated to $W. \to X_p$ in $HRR(X_p)$ by setting $(R\Gamma_p^{X.}(W.))_{s.}$ to be the rigid product of the rigid pull-backs of $W. \to X_p$ via the maps $X_s \to X_p$ indexed by $\Delta[s]_p$. Then $(R\Gamma_p^{X.}(W.))_{p.} \to X_p$ factors through $W. \to X_p$ so that $HRR(X.) \to HRR(X_p)$ is left final. ■

The proof of the following proposition makes frequent use of the uniqueness of maps between rigid hypercoverings to insure the validity of simplicial identities. The usefulness of this proposition will become apparent when we consider function complexes.

PROPOSITION 4.7. *Let* X. *be a locally noetherian simplicial scheme and let* n *be a positive integer. Then the natural strict map*

$$(X.\otimes\Delta[n])_{et} \to (X.)_{et}\times\Delta[n] \quad \textit{in} \quad \text{pro-(s. sets)}$$

is an isomorphism with inverse provided by a strict map $(X.)_{et}\times\Delta[n]\to (X.\otimes\Delta[n])_{et}$.

Proof. Because the coskeleton functor commutes with disjoint unions, $U..\otimes\Delta[n]\to X.\otimes\Delta[n]$ defined by $(U..\otimes\Delta[n])_{s,t} = U_{s,t}\otimes\Delta[n]_s$ is a rigid hypercovering of $X.\otimes\Delta[n]$ whenever U.. is a rigid hypercovering of X. . Therefore, the identification of $(X.)_{et}\otimes\Delta[n]$ with $(X.\otimes\Delta[n])_{et}\circ(?\otimes\Delta[n])$ determines the strict map $(X.\otimes\Delta[n])_{et}\to(X.)_{et}\times\Delta[n]$. To prove the proposition, it suffices to exhibit a right adjoint for $(?\otimes\Delta[n])$, $\rho: HRR(X.\otimes\Delta[n])\to HRR(X.)$, so that the adjunction transformation $(?\otimes\Delta[n])\circ\rho\to id$ determines a strict map from $(X.\otimes\Delta[n])_{et}\circ(?\otimes\Delta[n]) = (X.)_{et}\times\Delta[n]$ to $(X.\otimes\Delta[n])_{et}$ and $(?\otimes\Delta[n])$ is left final.

For $U..\to X.\otimes\Delta[n]$ in $HRR(X.\otimes\Delta[n])$, we define $\rho(U..\to X.\otimes\Delta[n]) = U'_{..}\to X.$ in HRR(X.) by setting $U'_{k.}$ equal to the rigid product in $HRR(X_k)$ indexed by the $\binom{n+k}{k}$ non-degenerate simplices $<\sigma,\sigma'>\in(\Delta[k]\times\Delta[n])_{n+k}$ of $\sigma^*(U^{\sigma'}_{k+n.}\to X_{n+k}\otimes\{\sigma'\})$ in $HRR(X_k)$, where $X_{k+n}\otimes\{\sigma'\}$ is that copy of $X_{k+n}\otimes\Delta[n]_{k+n}$ indexed by σ' and where $U^{\sigma'}_{k+n.}$ is the restriction of $U_{k+n.}\to(X.\otimes\Delta[n])_{k+n}$ to $X_{k+n}\otimes\{\sigma'\}$. If $\alpha:\{0,\cdots,k\}\to\{0,\cdots,m\}$ is a non-decreasing map, we define the composite $pr_\sigma\circ\alpha: U'_{m.}\to U'_{k.}\to\sigma^*(U^{\sigma'}_{k+n.})$ by $\alpha'\circ pr_\mu: U'_{m.}\to\mu^*(U^{\mu'}_{m+n.})\to\sigma^*(U^{\sigma'}_{k+n.})$, where $<\mu,\mu'>\in(\Delta[m]\times\Delta[n])_{m+n}$ and $\alpha':\{0,\cdots,k+n\}\to\{0,\cdots,m+n\}$ are defined as follows: define α' by $\alpha'(j)=\alpha'(j-1)+1$ for j such that $\sigma(j)=\sigma(j-1)$

and $\alpha'(j) = \alpha'(j-1) + \alpha(\sigma(j)) - \alpha(\sigma(j-1))$ if $\sigma(j) \neq \sigma(j-1)$; define $\mu : \{0,\cdots,m+n\} \to \{0,\cdots,m\}$ to be the unique surjective, non-decreasing function such that $\mu(i) = \mu(i+1)$ whenever there exists a j such that $\alpha'(j) = i$ and $\sigma(j) = \sigma(j+1)$. We readily verify that μ and μ' are the unique surjective non-decreasing maps fitting in the commutative squares

$$\begin{array}{ccccc}
\Delta[n] & \xleftarrow{\sigma'} & \Delta[k+n] & \xrightarrow{\sigma} & \Delta[k] \\
\Big\downarrow \mathrm{id} & & \Big\downarrow \alpha' & & \Big\downarrow \alpha \\
\Delta[n] & \xleftarrow{\mu'} & \Delta[m+n] & \xrightarrow{\mu} & \Delta[m]
\end{array} .$$

Because these squares commute, $\alpha' : U_{m+n.} \to U_{k+n.}$ restricts to $\mu^*(U^{\mu'}_{m+n.}) \to \sigma^*(U^{\sigma'}_{k+n})$ covering $\alpha : X_m \to X_k$. Moreover, if $\beta : \{0,\cdots,m\} \to \{0,\cdots,p\}$ is another non-decreasing map, we readily check that $(\beta \circ \alpha)' = \beta' \circ \alpha'$. The fact that $U'_{..}$ is a well-defined bi-simplicial scheme now follows using rigidity.

We briefly sketch the proof of the adjointness of $? \otimes \Delta[n]$ and ρ. If $f : (W_{..} \otimes \Delta[n] \to X_. \otimes \Delta[n]) \to (U_{..} \to X_. \otimes \Delta[n])$ is a map in $\mathrm{HRR}(X_. \otimes \Delta[n])$, then we obtain $(W_{..} \to X_.) \to (\rho(U_{..}) \to X_.)$ in $\mathrm{HRR}(X_.)$ by defining $W_{k.} \to \rho(U_{..})_{k.} \to \sigma^*(U^{\sigma'}_{k+n.})$ as the restriction of $f_{k+n} : W_{k+n.} \otimes \{\sigma'\} \to U^{\sigma'}_{k+n.}$ to $\sigma^*(W_{k.})$. Conversely, if $g : (W_{..} \to X_.) \to (\rho(U_{..}) \to X_.)$ is a map in $\mathrm{HRR}(X_.)$, then we obtain $(W_{..} \otimes \Delta[n] \to X_. \otimes \Delta[n]) \to (U_{..} \to X_. \otimes \Delta[n])$ whose restriction to $W_{k.} \otimes \{\varepsilon\} \to U^{\varepsilon}_{k.}$ is defined by $\delta \circ pr_{\sigma} \circ g_k : W_{k.} \to \rho(U_{..})_{k.} \to \sigma^*(U^{\sigma'}_{k+n.}) \to U^{\varepsilon}_{k.}$, where $<\sigma,\sigma'>$ is some non-degenerate simplex of $(\Delta[k] \times \Delta[n])_{k+n}$ and where $\delta : \sigma^*(U^{\sigma'}_{k+n.}) \to U^{\varepsilon}_{k.}$ is induced by $\delta : U^{\sigma'}_{k+n.} \to U^{\varepsilon}_{k.}$ with $\delta : \{0,\cdots,k\} \to \{0,\cdots,k+n\}$ any strictly increasing map such that $\sigma \circ \delta = \mathrm{id}$ and $\delta : \Delta[n]_{k+n} \to \Delta[n]_k$ sends σ' to ε. ■

The following is an immediate corollary of Proposition 4.7.

COROLLARY 4.8. *Let* $f, g : X_. \to Y_.$ *be two maps of locally noetherian simplicial schemes related by a simplicial homotopy. Then* f_{et} *and* g_{et} *determine the same map in* pro-$\mathcal{H}$. ■

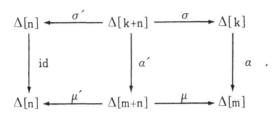

5. HOMOTOPY INVARIANTS

In this chapter, we consider the homotopy groups and cohomology groups of the etale topological type $(X.)_{et}$. We identify π_0 and π_1 with their algebraic counterparts (partial results about the higher homotopy groups are to be found in later chapters). In particular, our study of the fundamental group requires a review of descent techniques in the context of principal G-fibrations. Using our identification of fundamental groups, we also verify that the cohomology of $(X.)_{et}$ with abelian local coefficients is isomorphic to the cohomology of X. with coefficients in the corresponding locally constant sheaf. We point out that these homotopy and cohomology groups are invariants of the homotopy type of $(X.)_{et}$; by Theorem 3.8, the topological type itself determines in some sense the cohomology groups of X. with values in any abelian sheaf.

DEFINITION 5.1. Let X. ,x be a locally noetherian pointed simplicial scheme. For $n \geq 0$, we define the pro-object of pointed sets

$$\pi_n((X.\,,x)_{et}) = \pi_n \circ (X.\,,x)_{et} : \mathrm{HRR}(X.) \to (\mathrm{sets}).$$

Thus, for $n \geq 1$, $\pi_n((X.\,,x)_{et})$ is a pro-group; for $n \geq 2$, $\pi_n((X.\,,x)_{et})$ is a pro-abelian group. Let M be an abelian local coefficient system on $\pi \circ \Delta(U..)$ for some $U.. \in \mathrm{HRR}(X.)$ (see discussion preceding Corollary 5.8). For any $n \geq 0$, we define

$$H^n((X.)_{et}, M) = \mathrm{colim}\ H^n(\pi \circ \Delta(U'..), j^*M)$$

where the colimit is indexed by $j : U'.. \to U..$ in $\mathrm{HRR}(X.)/U..$. ■

The reason we consider the *homotopy pro-groups* rather than their inverse limits is that too much information is lost upon applying the inverse limit functor (which is not even exact, unlike the direct limit functor colim). As we see in the next proposition, $\pi_0((X.\ ,x))$ is actually (isomorphic to) a pointed set.

PROPOSITION 5.2. *Let* $X.\ ,x$ *be a locally noetherian pointed simplicial scheme and let* $\pi_0(X.\ ,x)$ *denote the pointed set of connected components of the pointed simplicial set* $\pi(X.\ ,x)$, $\pi_0(X.\ ,x) = \pi_0(\pi(X.\ ,x))$. *Then the pointed pro-set* $\pi_0((X.\ ,x)_{et})$ *is isomorphic in pro-(sets$_*$) to the pointed set* $\pi_0(X.\ ,x)$

$$\pi_0((X.\ ,x)_{et}) \simeq \pi_0(X.\ ,x)\ .$$

Furthermore, $X.$ *is a disjoint union of non-trivial simplicial schemes* $X.^a$ *corresponding to elements* $a \in \pi_0(X.\ ,x)$ *with no* $X.^a$ *expressible as a non-trivial disjoint union.*

Proof. We ignore the base point and prove that the natural map $\pi_0((X.)_{et}) \to \pi_0(X.)$ is an isomorphism of pro-sets. For this, it suffices to prove that this map induces a bijection

$$\mathrm{Hom}(\pi_0(X.),S) \to \mathrm{Hom}(\pi_0((X.)_{et},S) = \mathrm{colim}\ \mathrm{Hom}(\pi_0(\pi(\Delta U..)),S)$$

for any set S. To prove this, it suffices to prove that the natural map

$$H^0(X.\ ,S) = \mathrm{Hom}(\pi_0(X.),S) \to \mathrm{Hom}(\pi_0(\pi(\Delta U..)),S) = H^0(U..\ ,S)$$

is a bijection for any set S. Viewing S as a constant sheaf of sets, we may apply the proof of Proposition 3.7 to verify this last bijection. The second assertion of the proposition is verified by inspection. ∎

Grothendieck's algebraic interpretation of fundamental groups is given in terms of covering spaces. Those connected covering spaces which correspond to normal subgroups of the fundamental group are principal G-fibrations, where G is the quotient group.

DEFINITION 5.3. Let $X.$ be a simplicial scheme and let G be a (discrete) group. A *principal* G*-fibration* over $X.$ is a map of simplicial schemes $f: X'_. \to X.$ together with a right action of G on $X'_.$ over $X.$ such that

a.) There exists an etale, surjective map $U \to X_0$ and an isomorphism of schemes $U \underset{X_0}{\times} X'_0 \to U \otimes G$ over U commuting with the action of G (where G acts on $U \otimes G$ by right multiplication).

b.) For each map $\alpha: \Delta[n] \to \Delta[m]$ in Δ, the square

$$\begin{array}{ccc} X'_m & \xrightarrow{\alpha} & X'_n \\ {\scriptstyle f_m}\big\downarrow & & \big\downarrow{\scriptstyle f_n} \\ X_m & \xrightarrow{\alpha} & X_n \end{array}$$

is cartesian. A map of principal G-fibrations over $X.$, $X''_. \to X'_.$, is an isomorphism of simplicial schemes over $X.$ commuting with the action of G. We denote the category of principal G-fibrations over $X.$ by $\Pi(X.,G)$. ∎

In the following lemma, we re-interpret the above category $\Pi(X.,G)$ in terms of the category $\Pi(X_0,G)$ plus "descent data." We intend the following discussion to be a first introduction to the technique of descent.

LEMMA 5.4. *Let* $X.$ *be a simplicial scheme and let* G *be a group. Then the category* $\Pi(X.,G)$ *of principal* G*-fibrations over* $X.$ *is equivalent to the category* $\langle \Pi(X_0,G); \text{d.data} \rangle$ *defined as follows. An object of* $\langle \Pi(X_0,G); \text{d.data} \rangle$ *is a principal* G*-fibration over* X_0, $q_0: X'_0 \to X_0$, *together with "descent data"—namely, an isomorphism* $\phi: d_0^*(q_0) \to d_1^*(q_0)$ *in* $\Pi(X_1,G)$ *satisfying* $d_1^*\phi = d_2^*\phi \circ d_0^*\phi$ *in* $\Pi(X_2,G)$. *A map from* $(q_0: X'_0 \to X_0; \phi)$ *to* $(r_0: X''_0 \to X_0; \psi)$ *is a map* $\theta: X'_0 \to X''_0$ *in* $\Pi(X_0,G)$ *satisfying the condition* $\psi \circ d_0^*\theta = d_1^*\theta \circ \phi$ *in* $\Pi(X_1,G)$.

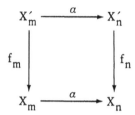

Proof. To a principal G-fibration $q: X'_. \to X_.$ over $X_.$, we associate the principal G-fibration $q_0: X'_0 \to X_0$ over X_0 together with the isomorphism ϕ given as the composite $d_0^*(q_0) \xrightarrow{\sim} q_1 \xrightarrow{\sim} d_1^*(q_0)$. Clearly, $q \mapsto <q_0, \phi>$ determines a faithful functor from $\Pi(X_., G)$ to $< \Pi(X_0, G); \text{d. data} >$.

Conversely, let $q_0: X'_0 \to X_0$ and $\phi: d_0^*(q_0) \to d_1^*(q_0)$ determine an element of $< \Pi(X_0, G); \text{d. data} >$. We define $q: X'_. \to X_.$ by setting X'_n equal to the fibre product of $d_0 \circ \cdots \circ d_0 : X_n \to X_0$ and $q_0: X'_0 \to X_0$ and by setting $q_n: X'_n \to X_n$ equal to the projection. Define $s_j: X'_{n-1} \to X'_n$ to be $s_j \times \mathrm{id}_{X'_0}$; define $d_i: X'_n \to X'_{n-1}$ to be $d_i \times \mathrm{id}_{X'_0}$ for $0 \le i < n$; and define $d_n: X'_n \to X'_{n-1}$ to be the composition of $(d_0 \circ \cdots \circ d_0)^* \phi$ (from X'_n to the fibre product of $d_1 \circ d_0 \circ \cdots \circ d_0 = d_0 \circ \cdots \circ d_0 \circ d_n : X_n \to X_0$ and $q_0: X'_0 \to X_0$) and $d_n \times \mathrm{id}_{X'_0}$. Using the simplicial identities, we easily verify that $q: X'_. \to X_.$ is a well-defined map of simplicial schemes. Moreover, using this explicit construction, we see that if (q_0, ϕ) and (r_0, ψ) are elements of $< \Pi(X_0, G); \text{d. data} >$, then a map $\theta: q_0 \to r_0$ of schemes over X_0 with $\psi \circ d_0^* \theta = d_1^* \theta \circ \phi$ determines a map of simplicial schemes $q \to r$. In particular, the G-action on q_0 determines a G-action on q, so that $q: X'_. \to X_.$ is an element of $\Pi(X_., G)$. The same argument implies that a map $(q_0, \phi) \to (r_0, \psi)$ in $< \Pi(X_0, G); \text{d. data} >$ determines a map $q \to r$ in $\Pi(X_., G)$. ∎

For $Y_{..}$ is a bi-simplicial scheme, we interpret the category $\Pi(\Delta Y_{..}, G)$ in a similar fashion.

PROPOSITION 5.5. *Let $Y_{..}$ be a bi-simplicial scheme and G a (discrete) group. Then category $\Pi(\Delta Y_{..}, G)$ is equivalent to the category $< \Pi(Y_{0.}, G); \text{d. data} >$ defined as follows. An object of $< \Pi(Y_{0.}, G); \text{d. data} >$ is a principal G-fibration $q: Y'_. \to Y_{0.}$ provided with an isomorphism $\phi: d_0^*(q) \xrightarrow{\sim} d_1^*(q)$ in $\Pi(Y_{1.}, G)$ satisfying $d_1^* \phi = d_2^* \phi \circ d_0^* \phi$ in $\Pi(Y_{2.}, G)$; a map $(q, \phi) \to (r, \psi)$ is a map $\theta: q \to r$ in $\Pi(Y_{0.}, G)$ satisfying the condition $\psi \circ d_0^* \theta = d_1^* \theta \circ \phi$.*

Proof. The natural map $Y_{0.} \to \Delta Y..$ determines via pull-back a functor $\Pi(\Delta Y..,G) \to \Pi(Y_{0.},G)$. If $X'_{\cdot} \to \Delta Y..$ is a principal G-fibration and $Y'_{\cdot} \to Y_{0.}$ the pull-back, then we define $\phi : d_0^* Y'_{\cdot} \to d_1^* Y'_{\cdot}$ in $\Pi(Y_{1.},G)$ as follows. Namely, ϕ is equivalent to a map $\phi_0 : d_0^* Y'_0 \to d_1^* Y'_0$ in $\Pi(Y_{1,0},G)$ commuting with descent data by Lemma 5.4; ϕ_0 is determined by the descent data for $X'_{\cdot} \to \Delta Y..$ as the following composition

$$d_0^{h*} Y'_0 = s_0^{v*} \circ d_0^{v*} \circ d_0^{h*} Y'_0 \xrightarrow{\simeq} s_0^{v*} X'_1 \xrightarrow{\simeq} s_0^{v*} \circ d_1^{v*} \circ d_1^{h*} Y'_0 = d_1^{h*} Y'_0$$

(where the superscript v refers to the internal simplicial structure of $Y_{n.}$ and the superscript h refers to the external structure). The fact that this construction determines an equivalence of categories is verified as in the proof of Lemma 5.4. ∎

As we see in the next proposition, *the fundamental pro-group* $\pi_1((X.,x)_{et})$ plays the same role for principal G-fibrations over X. as the fundamental group of a simplicial set does for principal G-fibrations over that simplicial set.

PROPOSITION 5.6. *Let* X., x *be a pointed, connected, locally noetherian simplicial scheme and let* G *be a group. Then the set* $\mathrm{Hom}(\pi_1((X.,x)_{et}),G)$ *is in natural one-to-one correspondence with the set of isomorphism classes of principal* G*-fibrations over* X. *pointed above* x. *Moreover, if* U.., u *is any hypercovering of* X. *pointed above* x, *then the set* $\mathrm{Hom}(\pi_1(\pi(\Delta U..),u),G)$ *is in natural one-to-one correspondence with the set of isomorphism classes of principal* G*-fibrations over* X. *pointed above* u *whose restriction to* $U_{0,0}$ *are trivial.*

Proof. A homomorphism $\pi_1(\pi(\Delta U..),u) \to G$ is equivalent to an isomorphism class of pointed principal G-fibrations of simplicial sets over $\pi(\Delta U..),u$. As argued in the proof of Proposition 5.5, this is equivalent to an isomorphism class of pointed principal G-fibrations $p : T,t \to \pi(U_{0.}),u$ together with descent data $\phi : d_0^*(p) \to d_1^*(p)$ satisfying $d_1^*\phi = d_2^*\phi \circ d_0^*\phi$.

Since $U_{s.}$ is a hypercovering of X_s for $s \geq 0$, the category of principal G-fibrations over $\pi(U_{s.})$ is equivalent to the category of principal G-fibrations over X_s whose restriction to $U_{s,0}$ is trivial ([8], 10.7). The second assertion of the proposition is now immediately implied by Lemma 5.4. To prove the first assertion, it suffices to recall that $\mathrm{Hom}(\pi_1((X.\,,x)_{et}),G) = \mathrm{colim}\ \mathrm{Hom}(\pi_1(\pi(\Delta U..),u),G)$ by definition and that if $X'_. \to X.$ is a principal G-fibration then there exists $U.. \to X$ with the property that $X'_0 \underset{X_0}{\times} U_{0,0} \to U_{0,0}$ is the trivial principal G-fibration (if $U \to X_0$ trivializes $X'_0 \to X_0$, define $U..$ to be the "rigid nerve" of the rigid covering $R\Gamma_0^{X.}(U) \to X.$; cf. proof of Corollary 4.6). ■

Proposition 5.6 establishes the following useful fact which we state as a separate corollary. We implicitly use the fact that a map of pro-groups $\{H_i\,;\ i \in I\} \to \{G_j\,;\ j \in J\}$ which induces a bijection $\mathrm{Hom}(\{G_j\},K) \xrightarrow{\simeq} \mathrm{Hom}(\{H_i\},K)$ for all groups K is necessarily an isomorphism of pro-groups.

COROLLARY 5.7. *Let* $X.\,,x$ *be a pointed, connected, locally noetherian simplicial scheme. Let* C *be any left filtering subcategory of* $HR(X.\,,x)$ *(the homotopy category of pointed hypercoverings of* $X.\,,x$ *) with the property that for any etale surjective map* $U \to X_0$ *there exists some* $U.. \to X.$ *in* C *such that* $U_{0,0}$ *factors through* $U \to X_0$. *Then the functors* $C \to HR(X.\,,x) \leftarrow HRR(X.)$ *induce isomorphisms of pro-groups*

$$\{\pi_1(\pi(\Delta U..),u);\ U..,\ u \in C\} \leftarrow \{\pi_1(\pi(\Delta V..),v);\ V..\ ,v \in HR(X.\,,x)\} \to \pi_1((X.\,,x)_{et}).\ ■$$

The reader should observe that $HR(X.\,,x)$ is indeed left filtering. Namely, if $f,g : U..\,,u \rightrightarrows V..\,,v$ are pointed maps of hypercoverings, then the left equalizer $h : H.. \to U..$ of f,g is naturally pointed by $(v \otimes 1, u)$: $\mathrm{Spec}\ \Omega \to H_{0,0}$, where $v \otimes 1 : \mathrm{Spec}\ \Omega \to \mathcal{H}om(\Delta[1],V_{0.})_0$ is the constant homotopy.

If $T.$ is a simplicial set, then a *local coefficient system* on $T.$ is a functor

$$L : (\Delta/T.)^0 \to (\text{sets})$$

assigning to each simplex $t_m \in T_m$ a set $L(t_m)$ and to each $\alpha : \Delta[n] \to \Delta[m]$ in Δ an isomorphism $L(\alpha) : L(\alpha(t_m)) \to L(t_n)$ (cf. [42] for a discussion of $\Delta/T.$, the category of simplices of $T.$). A map of local coefficient systems is a natural transformation of functors. An isomorphism class of local coefficient systems isomorphic to L with fibres isomorphic to S on the pointed, connected simplicial set $T.$, t naturally corresponds to an equivalence class of homomorphisms

$$\pi_1(T.,t) \to \mathrm{Aut}(S) ,$$

where the equivalence relation is generated by inner automorphisms of Aut(S). Such an equivalence class of homomorphisms naturally corresponds to an isomorphism class of principal Aut(S)-fibrations on $T.$.

For $\{T^i_. ; i \in I\} \in$ pro-(s. sets), a local coefficient system on $\{T^i_.\}$ is an equivalence class of local coefficient systems on some $T^j_.$, where $L^i_.$ on $T^i_.$ and $L^j_.$ on $T^j_.$ are equivalent if there exists some k mapping to i and j in I such that the restrictions of $L^i_.$ and $L^j_.$ to $T^k_.$ are isomorphic. Because the restrictions of a local coefficient system L on a simplicial set $T.$ via weakly homotopic maps $S. \rightrightarrows T.$ are isomorphic, this definition also applies to $\{T^i_.\} \in \text{pro-}\mathcal{H}$, where $\mathcal{H}$ is the homotopy category of simplicial sets. As discussed above, if $\{T^i_.\} \in \text{pro-(s. sets}_{*c})$ is a pro-object of pointed, connected simplicial sets, then an isomorphism class of local coefficient systems on $\{T^i_.\}$ with fibres isomorphic to a given set S is equivalent to an equivalence class of homomorphisms $\pi_1(\{T^i_.\}) = \{\pi_1(T^i_.)\} \to \mathrm{Aut}(S)$.

With these preliminaries, we give the following corollary of Proposition 5.6.

COROLLARY 5.8. *Let* $X.$ *be a connected, locally noetherian simplicial scheme, let* $U.. \to X.$ *be a hypercovering, and let* S *be a set. Then there is a natural one-to-one correspondence between the set of isomorphism classes of locally constant sheaves on* Et(X.) *with stalks isomorphic to* S *whose restrictions to* $U_{0,0}$ *are constant and the set of isomorphism classes of local coefficient systems on* $\pi(\Delta U..)$ *with fibres isomorphic to* S. *In particular, an isomorphism class of locally constant sheaves on* Et(X.) *is in natural one-to-one correspondence with an isomorphism class of local coefficient systems on* $(X.)_{et}$ *which is in natural one-to-one correspondence with an equivalence class of homomorphisms* $\pi_1((X.\,,x)_{et}) \to \mathrm{Aut}(S)$ *for any geometric point* x *of* X_0, *where* S *is isomorphic to the stalks.*

Proof. As seen above, a local coefficient system L on $\pi(\Delta U..)$ with stalks isomorphic to S is naturally equivalent to a principal Aut(S)-fibration over $\pi(\Delta U..)$. By Proposition 5.6, this is naturally equivalent to a principal Aut(S)-fibration over X. whose restriction to $U_{0,0}$ is trivial. Such a principal Aut(S)-fibration is easily seen to be equivalent to a map of simplicial schemes $X'. \to X.$ satisfying 5.3.b.) such that $X'_0 \underset{X_0}{\times} U_{0,0} \simeq U_{0,0} \otimes S$. Such a map $X'. \to X.$ is equivalent to the locally constant sheaf of sets on Et(X.) with stalks isomorphic to S defined by sending $U \to X_m$ to the set of maps from U to X'_m over X_m. This proves the first asserted equivalence of categories; the second follows from this. ∎

Using Corollary 5.8, we now identify the cohomology of $(X.)_{et}$ with coefficients in an abelian local coefficient system.

PROPOSITION 5.9. *Let* X. *be a connected, locally noetherian simplicial scheme, let* M *be a locally constant abelian sheaf on* Et(X.), *and let* M *also denote the corresponding abelian local coefficient system on* $(X.)_{et}$. *Then there is a natural isomorphism*

$$H^*(X.\,,M) \simeq H^*((X.)_{et},M)\ .$$

Proof. If L is an abelian local coefficient system on a simplicial set S., then $H^*(S.\,,L)$ is defined to be the cohomology of the complex

$\{n \mapsto \prod_{s_n \in S_n} L(s_n)\}$. If $U.. \to X.$ is a hypercovering of $X.$ such that the locally constant abelian sheaf M is constant when restricted to $U_{0,0}$, then the cohomology of the bi-complex $M(U..)$ is naturally isomorphic to $H^*(\pi(\Delta U..),M)$. Thus, the proposition follows from Theorem 3.8 by taking colimits indexed by the left cofinal subcategory of $HRR(X.)$ consisting of $U.. \to X.$ such that M restricted to $U_{0,0}$ is constant. ∎

We observe that the hypothesis that $X.$ be connected in Proposition 5.9 is not necessary because we can use Proposition 5.2 to equate $H^*(X.,M)$ with $\prod_{\alpha \in \pi_0(X.)} H^*(X^\alpha.,M)$ and $H^*((X.)_{et},M)$ with $\prod_{\alpha \in \pi_0(X.)} H^*((X^\alpha_.)_{et},M)$.

6. WEAK EQUIVALENCES, COMPLETIONS, AND HOMOTOPY LIMITS

In this chapter, we review the homotopy theory which has been employed in various applications of the etale topological type. The role of the constructions we consider is to enable us to obtain homotopy theoretic information from the invariants of $(X.)_{et}$ considered in Chapter 5.

After reviewing the definitions of various homotopy categories, we present the theorem of M. Artin and B. Mazur which provides necessary and sufficient conditions for a map to be a weak equivalence in the pro-homotopy category [8]. We then proceed to recall the Artin-Mazur pro-L completion functor which enables us to exclude from consideration homotopy information at specified primes. Following D. Sullivan, we next consider the Sullivan homotopy limit $\underleftarrow{\text{holim}}^{Su}(\ )$ which provides a categorical inverse limit for certain pro-objects in the homotopy category [69]. Finally, we consider the constructions $(\mathbb{Z}/\ell)_\infty(\ )$ and $\underleftarrow{\text{holim}}(\ )$ of A. K. Bousfield and D. M. Kan [13]. These ℓ-completion and homotopy limit functors have the significant advantage of being "rigid" in the sense that they take values in $(s.\ sets_*)$ rather than in the homotopy category. As we see, these constructions often provide a rigid version of the homotopy theoretic constructions of Artin, Mazur, and Sullivan.

We recall that a map $f: S. \to T.$ in $(s.\ sets)$ is said to be a *weak equivalence* if the geometric realization of f is a homotopy equivalence. We let $\mathcal{H}$, the *homotopy category*, denote the category obtained from $(s.\ sets)$ by formally inverting the weak equivalences. Similarly, $\mathcal{H}_*$ is obtained from $(s.\ sets_*)$ by inverting pointed maps which are weak equivalences. We let $(s.\ sets_{*c})$ denote the category of pointed, connected simplicial sets and $\mathcal{H}_{*c}$ denote its homotopy category (a full subcategory of $\mathcal{H}_*$). Thus, a map in (s. sets$_{*c}$) is invertible in $\mathcal{H}_{*c}$ if and only if it induces an isomorphism on homotopy groups. Although we shall not require their construction, we also mention the construction of Ho(pro-(s. sets$_*$)), a homotopy category of pro-simplicial sets, by D. A. Edwards and H. M. Hastings [28].

We recall that the n-th coskeleton functor,

$$\operatorname{cosk}_n : (\text{s. sets}_*) \to (\text{s. sets}_*) ,$$

is the right adjoint of $\mathrm{sk}_n(\ )$, the n-skeleton functor (as in Definition 1.3). Unlike $\mathrm{sk}_n(\ )$, $\operatorname{cosk}_n(\ )$ induces a functor $\operatorname{cosk}_n : \mathcal{H}_* \to \mathcal{H}_*$, because for any pointed, connected simplicial set $T.$ the natural map $T. \to \operatorname{cosk}_n(T.)$ induces isomorphisms $\pi_k(T.) \to \pi_k(\operatorname{cosk}_n T.)$ for $k < n$ and the homotopy groups $\pi_k(\operatorname{cosk}_n T.)$ are zero for $k \geq n$. This observation permits us to give the following definition.

DEFINITION 6.1. The functor $\# : (\text{s. sets}_*) \to (\text{pro-s. sets}_*)$ sending $T.$ to $\#(T.) = \{\operatorname{cosk}_n T. \, ; \, n \geq 0\}$ extends to a functor

$$\# : \text{pro-}\mathcal{H}_* \to \text{pro-}\mathcal{H}_*$$

by sending $\{S^i_. \, ; \, i \in I\}$ to $\{\operatorname{cosk}_n S^i_. \, ; \, (n,i) \in \mathbf{N} \times I\}$. A map $f : \{S^i_.\} \to \{T^j_.\}$ in pro-$\mathcal{H}_*$ is said to be a *weak equivalence in pro-*$\mathcal{H}_*$ if $\#(f) : \#\{S^i_.\} \to \#\{T^j_.\}$ is an isomorphism in pro-$\mathcal{H}_*$. ■

The introduction of weak equivalences in pro-$\mathcal{H}_*$ is justified by the following theorem. As in Chapter 5 when we considered objects in pro-(s. sets$_{*c}$), for any $\{S^i_.\} \in$ pro-$\mathcal{H}_{*c}$ we define $\pi_k(\{S^i_.\}) = \{\pi_k(S^i_.)\}$ and $H^*(\{S^i_.\},M) = \operatorname{colim} H^*(S^i_. ,M)$ for any abelian local coefficient system M on $\{S^i_.\}$ (given by an equivalence class of homomorphisms $\pi_1(\{S^i_.\}) \to \operatorname{Aut}(M_0)$, where M_0 is a stalk of M; the colimit is indexed by the left final full subcategory of I consisting of those j for which $\pi_1(\{S^i_.\}) \to \operatorname{Aut}(M_0)$ factors through $\pi_1(\{S^i_.\}) \to \pi_1(S^j_.)$).

THEOREM 6.2 (M. Artin and B. Mazur [8], 4.3 and 4.4). *For any map* $f:\{S^i_.\} \to \{T^j_.\}$ *in* pro-$\mathcal{H}_{*c}$, *the following are equivalent.*

a.) f *is a weak equivalence in* pro-$\mathcal{H}_*$.

b.) $f_*: \pi_k(\{S^i_.\}) \to \pi_k(\{T^j_.\})$ *is an isomorphism for all* $k > 0$.

c.) $f_*: \pi_1(\{S^i_.\}) \to \pi_1(\{T^j_.\})$ *is an isomorphism; and for every abelian local coefficient system* M *on* $\{T^j_.\}$, $f^*: H^*(\{T^j_.\},M) \to H^*(\{S^i_.\},M)$ *is an isomorphism.* ∎

We observed in Chapter 5 that the fundamental pro-group and the cohomology groups with abelian local coefficients of $(X_.\,,x)_{et}$ can be identified "algebraically." Theorem 6.2 implies that these algebraic invariants determine $(X_.\,,x)_{et}$ up to weak equivalence in pro-$\mathcal{H}_*$. As a first application of Theorem 6.2 in conjunction with Chapter 5, we provide the following corollary.

COROLLARY 6.3. *Let* $X_.\,,x$ *be a pointed, connected, locally noetherian simplicial scheme and let* $HR(X_.\,,x)$ *be as in Corollary 5.7. Define* $(X_.\,,x)_{ht} \in$ pro-$\mathcal{H}_{*c}$ *by*

$$(X_.\,,x)_{ht} = \pi \circ \Delta(\ ): HR(X_.\,,x) \to \mathcal{H}_{*c}\,.$$

Then the natural map $HRR(X_.) \to HR(X_.\,,x)$ *determines a weak equivalence in* pro-$\mathcal{H}_*$, $(X_.\,,x)_{ht} \to (X_.\,,x)_{et}$.

Proof. We verify that the forgetful functor $HR(X_.\,,x) \to HR(X_.)$ is left final. Let $U_{..}\,,u$ be a pointed hypercovering of $X_.\,,x$ and let $f,g: U_{..} \to V_{..}$ be two maps of hypercoverings of $X_.$. Then $f_*(u), g_*(u): \mathrm{Spec}\,\Omega \to V_{0,0}$ extend to $k: \mathrm{Spec}\,\Omega \otimes \Delta[1] \to V_{0.}$ over the constant homotopy $\mathrm{Spec}\,\Omega \otimes \Delta[1] \to X_0$ because $V_{0.} \times_{X_0} \mathrm{Spec}\,\Omega$ is contractible (see the proof of Proposition 3.6). Thus $(k,u): \mathrm{Spec}\,\Omega \to H_{..}$ is a pointing of the left equalizer $h: H_{..} \to U_{..}$ of f,g with $h_*(k,u) = u$.

By Corollary 5.7, $j:(X_.\,,x)_{ht} \to (X_.\,,x)_{et}$ in pro-$\mathcal{H}_{*c}$ induces an isomorphism on fundamental pro-groups. By Theorem 6.2, it suffices to verify that j induces an isomorphism $j^*: H^*((X.\ ,x)_{et},M) \to H^*((X.\ ,x)_{ht},M)$ for any abelian local coefficient system M on $(X.\ ,x)_{et}$. Using the cofinality of $HR(X.\ ,x) \to HR(X.)$, we identify j^* with the map

$$\operatorname*{colim}_{HRR(X.)} H^*(\pi(\Delta U..),M) \to \operatorname*{colim}_{HR(X.)} H^*(\pi(\Delta V..),M)$$

which was proved to be an isomorphism in Corollary 4.6. ∎

If G is a group and L a set of primes, the *pro-L completion* of G is the system of finite, L-torsion quotient groups (i.e., only primes in L divide their orders) of G. By choosing a small left final subcategory I of the category of all such quotient homomorphisms, we obtain a pro-group $(G)^{\hat{L}}: I \to$ (finite L-groups). As is easy to check, this construction extends to a functor

$$(\ )^{\hat{L}}: \text{pro-(groups)} \to \text{pro-(finite L-groups)}\ .$$

The following theorem of M. Artin and B. Mazur provides the analagous construction for pro-$\mathcal{H}_{*c}$.

THEOREM 6.4 (M. Artin and B. Mazur [8], 3.4 and 4.3). *Let* L *be a set of primes. Then the inclusion* pro-$L\mathcal{H}_{*c} \to$ pro-$\mathcal{H}_{*c}$ *has a left adjoint*

$$(\ )^{\hat{L}}: \text{pro-}\mathcal{H}_{*c} \to \text{pro-}L\,\mathcal{H}_{*c}$$

where $L\mathcal{H}_{*c}$ *is the full subcategory of* $\mathcal{H}_{*c}$ *consisting of simplicial sets whose homotopy groups are finite* L*-groups. The map on fundamental pro-groups induced by the canonical map* $\{S^i_.\} \to (\{S^i_.\})^{\hat{L}}$ *is the* pro-L *completion map for* $\pi_1(\{S^i_.\})$ *for any* $\{S^i_.\} \in$ pro-$\mathcal{H}_{*c}$. *Moreover, if* M *is any abelian local coefficient system on* $(\{S^i_.\})^{\hat{L}}$ *whose fibres are finite* L*-groups, then the canonical map induces an isomorphism* $H^*((\{S^i_.\})^{\hat{L}},M) \xrightarrow{\sim} H^*(\{S^i_.\},M)$. ∎

As a corollary of Theorems 6.2 and 6.4, we conclude the following.

COROLLARY 6.5. *Let* $f:\{S^i_.\}\to\{T^j_.\}$ *be a map in* pro-$\mathcal{H}_{*c}$ *and let* L *be a set of primes. Then* $(f)^{\hat{L}}$ *is a weak equivalence in* pro-$\mathcal{H}_*$ *if and only if*

a.) $(f_*)^{\hat{L}}:(\pi_1(\{S^i_.\}))^{\hat{L}}\to(\pi_1(\{T^j_.\}))^{\hat{L}}$ *is an isomorphism; and*

b.) *for every abelian local coefficient system* M *on* $\{T^j_.\}$ *whose fibres are finite* L*-groups and which is represented by a map* $\pi_1(\{T^j_.\})\to \mathrm{Aut}(M_0)$ *factoring through* $\pi_1(\{T^j_.\})\to(\pi_1(\{T^j_.\}))^{\hat{L}}$, $f^*:H^*(\{T^j_.\},M)\to H^*(\{S^i_.\},f^*M)$ *is an isomorphism.*

Proof. The validity of a.) and b.) whenever $(f)^{\hat{L}}$ is a weak equivalence in pro-$\mathcal{H}_*$ is immediate from Theorem 6.4. Conversely, any abelian local coefficient system M on $\{T^j_.\}$ has a sub-system M_L whose fibres are the maximal L-torsion subgroups of the fibres of M. Furthermore, for any $k>0$, $H^k((\{T^j_.\})^{\hat{L}},M_L)\xrightarrow{\sim}H^k((\{T^j_.\})^{\hat{L}},M)$ because $(\{T^j_.\})^{\hat{L}}$ is in pro-L$\mathcal{H}_{*c}$. Because M_L on $(\{T^j_.\})^{\hat{L}}$ is a colimit of local coefficient systems with finite L-groups as fibres and because $H^*((\{T^j_.\})^{\hat{L}},\ )$ commutes with directed colimits, we conclude using Theorems 6.2 and 6.4 that a.) and b.) imply that $(f)^{\hat{L}}$ is a weak equivalence in pro-$\mathcal{H}_*$. ■

We now consider the *Sullivan homotopy limit* functor, $\operatorname{holim}\limits_{\leftarrow}{}^{Su}(\ )$, which enables us to associate a single homotopy type to $((X.)_{et})^{\hat{L}}$ in pro-L$\mathcal{H}_{*c}$.

THEOREM 6.6 (D. Sullivan [69], 3.1). *Let* P *denote the set of all primes. There exists a functor*

$$\operatorname{holim}\limits_{\leftarrow}{}^{Su}(\ ):\text{pro-P}\mathcal{H}_{*c}\to\mathcal{H}_{*c}$$

together with a natural transformation $\operatorname{holim}\limits_{\leftarrow}{}^{Su}(\ )\to \mathrm{id}:\text{pro-P}\mathcal{H}_{*c}\to\text{pro-}\mathcal{H}_{*c}$ *characterized by the property that this natural transformation induces a bijection*

$$\mathrm{Hom}_{\mathcal{H}_*}(S.\ ,\operatorname{holim}\limits_{\leftarrow}{}^{Su}(\{T^j_.\}))\to\mathrm{Hom}_{\text{pro-}\mathcal{H}_*}(S.\ ,\{T^j_.\})=\lim_{\substack{\leftarrow\\ J}}\mathrm{Hom}_{\mathcal{H}_*}(S.\ ,T^j_.)$$

for every S. *in* $\mathcal{H}_{*c}$ *and* $\{T^j_\cdot\,;\, j \in J\} \in$ pro-$P\mathcal{H}_{*c}$. *We choose* $\underleftarrow{\mathrm{holim}}^{Su}(\{T^j_\cdot\})$ *to be a Kan complex for each* $\{T^j_\cdot\} \in$ pro-$P\mathcal{H}_{*c}$. *Moreover, for any set of primes* L *the Sullivan pro-*L *completion functor*

$$\underleftarrow{\mathrm{holim}}^{Su}(\ )\circ(\ )^{\hat{L}} : \text{pro-}\mathcal{H}_{*c} \to \mathcal{H}_{*c}$$

has the property that if T. *is a pointed, connected simplicial set with fundamental group solvable of finite type and finitely generated higher homotopy groups, then the canonical map* $T_\cdot \to \underleftarrow{\mathrm{holim}}^{Su}((T_\cdot)^{\hat{L}})$ *induces an isomorphism in cohomology*

$$H^*(\underleftarrow{\mathrm{holim}}^{Su}((T_\cdot)^{\hat{L}}),A) \xrightarrow{\sim} H^*(T_\cdot\,,A)$$

for any finite abelian L*-group* A *and isomorphisms*

$$\pi_i(\underleftarrow{\mathrm{holim}}^{Su}((T_\cdot)^{\hat{L}})) \xrightarrow{\sim} \underleftarrow{\lim}((\pi_i(T_\cdot)^{\hat{L}}) \qquad i > 0\ . \ \blacksquare$$

Using the characterizing property of $\underleftarrow{\mathrm{holim}}^{Su}(\ )$, we immediately conclude the isomorphism $\pi_*(\underleftarrow{\mathrm{holim}}^{Su}(\{T^j_\cdot\})) \xrightarrow{\sim} \underset{J}{\underleftarrow{\lim}}\, \pi_*(T^j_\cdot)$ for any $\{T^j_\cdot\,;\, j \in J\}$ in pro-$P\mathcal{H}_{*c}$; in particular, $\pi_i(\underleftarrow{\mathrm{holim}}^{Su}((X_\cdot\,,x)_{et}{}^{\hat{L}})) = \underleftarrow{\lim}(\pi_i(X_\cdot\,,x)_{et}{}^{\hat{L}})$. Unfortunately, Theorem 6.6 does not in general enable us to determine either the higher homotopy groups or cohomology groups of $\underleftarrow{\mathrm{holim}}^{Su}((X_\cdot\,,x)_{et}{}^{\hat{L}})$. Nevertheless, the following immediate corollary of Corollary 6.5 and the isomorphism $\pi_*(\underleftarrow{\mathrm{holim}}^{Su}(\ )) \xrightarrow{\sim} \underleftarrow{\lim}\,\pi_*(\ )$ is often very useful in determining $\underleftarrow{\mathrm{holim}}^{Su}((X_\cdot\,,x)_{et}{}^{\hat{L}})$.

COROLLARY 6.7. *Let* $f : \{S^i_\cdot\} \to \{T^j_\cdot\}$ *be a map in* pro-$\mathcal{H}_{*c}$ *satisfying* a.) *and* b.) *of Corollary 6.5 for some set* L *of primes. Then* f *induces a homotopy equivalence of Sullivan* pro-L *completions,*

$$\underleftarrow{\mathrm{holim}}^{Su}((\{S^i_\cdot\})^{\hat{L}}) \to \underleftarrow{\mathrm{holim}}^{Su}((\{T^j_\cdot\})^{\hat{L}})\ . \ \blacksquare$$

The constructions we have considered have not used the fact that $(X.\,,x)$ is an object of pro-(s. sets$_*$) rather than simply in pro-$\mathcal{H}_*$. We now consider constructions of A. K. Bousfield and D. M. Kan which utilize this extra structure. The Bousfield-Kan $\mathbb{Z}/\ell$-*completion* functor for some prime ℓ,

$$(\mathbb{Z}/\ell)_\infty(\ ) : (\text{s. sets}) \to (\text{s. sets}) ,$$

is a functor together with a natural transformation $\mathrm{id} \to (\mathbb{Z}/\ell)_\infty(\ )$ such that a map $f : S. \to T.$ induces an isomorphism $f^* : H^*(T.\,, \mathbb{Z}/\ell) \xrightarrow{\sim} H^*(S.\,, \mathbb{Z}/\ell)$ if and only if $(\mathbb{Z}/\ell)_\infty(f)$ is a homotopy equivalence. Moreover, $(\mathbb{Z}/\ell)_\infty(f)$ is a Kan fibration whenever $f : S. \to T.$ is surjective; because $(\mathbb{Z}/\ell)_\infty(*) = *$, where $*$ is the trivial pointed simplicial set, $(\mathbb{Z}/\ell)_\infty(S.)$ is a Kan complex for any simplicial set $S.$ and $(\mathbb{Z}/\ell)_\infty(\ )$ determines $(\mathbb{Z}/\ell)_\infty(\ ) : (\text{s. sets}_*) \to (\text{s. sets}_*)$ ([13], I.4.2).

The Bousfield-Kan *homotopy* (inverse) *limit* functor

$$\underleftarrow{\mathrm{holim}}(\ ) : (\text{s. sets}^I) \to (\text{s. sets})$$

is a functor on the category, (s. sets^I), of "I-diagrams" of simplicial sets for any small category I (i.e., functors from I to (s. sets)). The naturality with respect to I enables one to extend $\mathrm{holim}(\ )$ to be functorial with respect to strict maps: if $\alpha : I \to J$, $G : I \to (\text{s. sets})$, and $F : J \to (\text{s. sets})$ are functors, then a natural transformation $F \circ \alpha \to G$ determines a map $\underleftarrow{\mathrm{holim}}(F) \to \underleftarrow{\mathrm{holim}}(G)$. A natural transformation $F \to G$ of I-diagrams with the property that $F(i) \to G(i)$ is a homotopy equivalence for each $i \in I$ determines a homotopy equivalence $\underleftarrow{\mathrm{holim}}(F) \to \underleftarrow{\mathrm{holim}}(G)$ provided that each $F(i)$ and $G(i)$ are Kan complexes. Moreover, if $\alpha : I \to J$ is a left final functor between small left filtering categories and $G : J \to (\text{s. sets})$ is such that $G(j)$ is a Kan complex for each $j \in J$, then $\underleftarrow{\mathrm{holim}}(G) \to \underleftarrow{\mathrm{holim}}(G \circ \alpha)$ is a homotopy equivalence ([13], XI.9.2). The homotopy limit of Kan complexes is again a Kan complex.

In the following theorem, we summarize several more useful properties of $(Z/\ell)_\infty(\ )$ and $\underleftarrow{\text{holim}}(\ )$. We recall that a *nilpotent simplicial set* is a pointed, connected simplicial set with nilpotent fundamental group which acts nilpotently on the higher homotopy groups.

THEOREM 6.8 (A. K. Bousfield and D. M. Kan [13], I.7.2, VI.6.2, VII.5.2, XI.7.1). *For any pair of simplicial sets* S. *and* T. *and any prime* ℓ, *the canonical map* $(Z/\ell)_\infty(S. \times T.) \to (Z/\ell)_\infty(S.) \times (Z/\ell)_\infty(T.)$ *is a homotopy equivalence with a natural right inverse.*

If $S. \in (\text{s. sets}_{*c})$ *either is nilpotent or has finite fundamental group, then* $S. \to (Z/\ell)_\infty(S.)$ *induces an isomorphism* $H^*((Z/\ell)_\infty(S.), Z/\ell) \to H^*(S., Z/\ell)$. *For any I-diagram* $F: I \to (\text{s. sets}_*)$ *with the property that* F(i) *is a Kan complex for each* $i \in I$, *there exists a spectral sequence of cohomological type*

$$E_2^{s,t} = \underleftarrow{\lim}^s\{\pi_t(F_i)\} \Rightarrow \pi_{t-s}(\underleftarrow{\text{holim}}(F)) \qquad 0 \le s \le t$$

(*where* $\underleftarrow{\lim}^0(\ )$ *is the inverse limit functor,* $\underleftarrow{\lim}^1(\ ): (\text{grps}^I) \to \text{grps})$ *its first derived functor, and* $\underleftarrow{\lim}^s(\ ): (\text{ab. grps}^I) \to (\text{ab. grps})$ *its* s-*th derived functor*) *which converges in positive degrees provided that* $\underset{r}{\underleftarrow{\lim}}^1 E_r^{s,t} = 0$ *for all* s,t *with* $0 \le s \le t$. ∎

The following proposition will usually be applied in the special case that each component of each $S.^i$ and $T.^j$ has finite homotopy groups (in which case $\underleftarrow{\lim}^k \pi_m(\{{}^\alpha S.^i\}) = 0$ for all $k,\ m \ge 1$). We require the generality of disconnected $S.^i$ and $T.^j$ for our discussion of function complexes in Chapter 11.

PROPOSITION 6.9. *Let* $f: \{S.^i\ ;\ i \in I\} \to \{T.^j\ ;\ j \in J\}$ *be a strict map of* pro-(s. sets$_*$) *which is a weak equivalence in* pro-$\mathcal{H}_*$. *Assume that each* $S.^i$, $T.^j$ *is a Kan complex. Then*

$$\underleftarrow{\text{holim}}(f): \underleftarrow{\text{holim}}\{S.^i\} \to \underleftarrow{\text{holim}}\{T.^j\}$$

restricts to a homotopy equivalence between distinguished connected components. Moreover, let $\pi = \underleftarrow{\lim}\, \pi_0(\{S^i_\cdot\}) = \underleftarrow{\lim}\, \pi_0(\{T^j_\cdot\})$ *and, for each* $\alpha \in \pi$, *let* ${}^{\alpha}S^i_\cdot$, ${}^{\alpha}T^j_\cdot$ *denote the corresponding connected components of* $S^i_\cdot$, $T^j_\cdot$. *Then* $\underleftarrow{\mathrm{holim}}(f)$ *is the disjoint union of maps* $\underleftarrow{\mathrm{holim}}({}^{\alpha}f): \underleftarrow{\mathrm{holim}}\{{}^{\alpha}S^i_\cdot\} \to \underleftarrow{\mathrm{holim}}\{{}^{\alpha}T^j_\cdot\}$. *If* $\underleftarrow{\lim}^k\, \pi_k(\{{}^{\alpha}S^i_\cdot\}) = 0$ *for all* $k > 0$, *then* $\underleftarrow{\mathrm{holim}}({}^{\alpha}f)$ *is a homotopy equivalence; if, in addition,* $\underleftarrow{\lim}^k\, \pi_{k+1}(\{{}^{\alpha}S^i_\cdot\}) = 0$ *for all sufficiently large* k, *then* $\underleftarrow{\mathrm{holim}}\{{}^{\alpha}S^i_\cdot\}$ *and* $\underleftarrow{\mathrm{holim}}\{{}^{\alpha}T^j_\cdot\}$ *are connected.*

Proof. We employ the constructions of [13]. Let $X_n = \mathrm{Tot}_n(\underset{\sim}{\Pi}^*\{S^i_\cdot\})$ and $Y_n = \mathrm{Tot}_n(\underset{\sim}{\Pi}^*\{T^j_\cdot\})$, so that $\underleftarrow{\mathrm{holim}}\{S^i_\cdot\} = \underleftarrow{\lim}\{X_n\}$ and $\underleftarrow{\mathrm{holim}}\{T^j_\cdot\} = \underleftarrow{\lim}\{Y_n\}$. The strict map f induces a map from the "first derived homotopy sequences" of [13], IX.4.1 for $\{X_n\}$

$$\cdots \to \pi_2 X^{(1)}_{m-1} \to \underleftarrow{\lim}^m\, \pi_{m+1}(\{X_n\}) \to \pi_1 X^{(1)}_m \to \pi_1 X^{(1)}_{m-1} \to$$
$$\underleftarrow{\lim}^m\, \pi_m(\{X_n\}) \to \pi_0 X^{(1)}_m \to \pi_0 X^{(1)}_{m-1} \qquad m \geq 0$$

to those for $\{Y_n\}$, where $\pi_i X^{(1)}_m = \mathrm{image}\,(\pi_i(X_{m+1}) \to \pi_i(X_m))$. Using [13], XI.7.1 and the isomorphism $\underleftarrow{\lim}^s\, \pi_t(\{S^i_\cdot\}) \xrightarrow{\sim} \underleftarrow{\lim}^s\, \pi_t(\{T^j_\cdot\})$, we conclude that f induces an isomorphism $\pi_i(X_n) \xrightarrow{\sim} \pi_i(Y_n)$ for each n and each $i > 0$. Consequently, the Milnor exact sequence implies that f induces isomorphisms $\pi_i(\underleftarrow{\lim}\, X_n) \xrightarrow{\sim} \pi_i(\underleftarrow{\lim}\, Y_n)$ for $i > 0$, so that $\underleftarrow{\mathrm{holim}}(f)$ restricts to an equivalence on connected components.

The fact that $\underleftarrow{\mathrm{holim}}(f) = \coprod_{\alpha \in \pi} \underleftarrow{\mathrm{holim}}({}^{\alpha}f)$ is immediate from the observation that any map $\underset{\sim}{\Delta} \to \underset{\sim}{\Pi}^*\{S^i_\cdot\}$ factors through $\underset{\sim}{\Pi}^*\{{}^{\alpha}S^i_\cdot\} \to \underset{\sim}{\Pi}^*\{S^i_\cdot\}$ for some α. Fix some $\alpha \in \pi$, and define ${}'S^i_\cdot = \underset{i/I}{\underleftarrow{\mathrm{holim}}}\{{}^{\alpha}S^k_\cdot\}$ and ${}'T^j_\cdot = \underset{j/J}{\underleftarrow{\mathrm{holim}}}\{{}^{\alpha}T^{\ell}_\cdot\}$.

Then the natural maps ${}^{\alpha}S^i_\cdot \to {}'S^i_\cdot$ and ${}^{\alpha}T^j_\cdot \to {}'T^j_\cdot$ are homotopy equivalences ([13], XI.4.1), so that $\underleftarrow{\mathrm{holim}}({}^{\alpha}f)$ is a homotopy equivalence if and only if $\underleftarrow{\mathrm{holim}}({}'f): \underleftarrow{\mathrm{holim}}\{{}'S^i_\cdot\} \to \underleftarrow{\mathrm{holim}}\{{}'T^j_\cdot\}$ is a homotopy equivalence.

Observe that a choice of base point of $\underleftarrow{\mathrm{holim}}\{{}^{\alpha}S^i_\cdot\}$ provides $'f:\{'S^i_\cdot\} \to \{'T^j_\cdot\}$ with the structure of a strict map in pro-(s. sets$_{*c}$) which is an isomorphism in pro-$\mathcal{H}_*$.

Consequently, $\underleftarrow{\mathrm{holim}}('f):\underleftarrow{\mathrm{holim}}\{'S^i_\cdot\} \to \underleftarrow{\mathrm{holim}}\{'T^j_\cdot\}$ restricts to a homotopy equivalence on connected components associated to any base point of $\underleftarrow{\mathrm{holim}}\{{}^{\alpha}S^i_\cdot\}$. If $\underleftarrow{\lim}^m \pi_m(\{{}^{\alpha}S^i_\cdot\}) = 0$ for all $m \geq 0$, then we conclude using the above first derived homotopy sequences that $\pi_0 {}'X^{(1)}_m = \{*\} = \pi_0 {}'Y^{(1)}_m$ for all $m \geq 0$. This implies that $\underleftarrow{\mathrm{holim}}(f)$ induces a bijection on connected components, with $\pi_0(\underleftarrow{\mathrm{holim}}\{'S^i_\cdot\}) \simeq \underleftarrow{\lim}^1 \pi_1{}'X_m \simeq \pi_0(\underleftarrow{\mathrm{holim}}\{'T^j_\cdot\})$; hence $\underleftarrow{\mathrm{holim}}('f)$ is a homotopy equivalence. If $\underleftarrow{\lim}^{k-1} \pi_k(\{{}^{\alpha}S^i_\cdot\}) = 0$ for k sufficiently large, then these first derived homotopy sequences imply that $\{\pi_1('X_n)\}$ is Mittag-Leffler so that $\underleftarrow{\mathrm{holim}}\{'S^i_\cdot\}$ and $\underleftarrow{\mathrm{holim}}\{'T^j_\cdot\}$ are connected. ■

We conclude this chapter by comparing the Sullivan and the Bousfield-Kan completions and homotopy limits. The proof we give is based in part on unpublished correspondence of A. K. Bousfield.

PROPOSITION 6.10. *Let ℓ be a prime and let $\ell\mathcal{H}_{*c}$ be the full subcategory of $\mathcal{H}_{*c}$ consisting of pointed, connected simplicial sets whose homotopy groups are finite ℓ-groups. For any* $T.$ *in $\mathcal{H}_{*c}$ with finite homotopy groups, there is a natural isomorphism in* pro-$\ell\mathcal{H}_{*c}$

$$(\mathbf{Z}/\ell)_\infty(T.) \to (T.)^{\widehat{\ell}}.$$

More generally, for any $T. \in \mathcal{H}_{*c}$ *with* $H^k(T.,\mathbf{Z}/\ell)$ *finite for each* $k > 0$, *there is a natural weak equivalence in* pro-$\ell\mathcal{H}_{*c}$

$$(T.)^{\widehat{\ell}} \to \{(\mathbf{Z}/\ell)_n(T.);\ n > 0\}$$

where $(\mathbf{Z}/\ell)_\infty(\ ) = \underleftarrow{\lim}(\mathbf{Z}/\ell)_n(\ )$. *For* $\{T^j_\cdot\} \in$ pro-(s. sets$_{*c}$) *such that* $\pi_k(\{T^j_\cdot\})$ *is pro-finite for each* $k > 0$ *and each* $T^j_\cdot$ *is a Kan complex, there is a natural isomorphism in* $\mathcal{H}_*$

$$\operatorname{holim}_{\leftarrow}\{T_.^j\} \to \operatorname{holim}_{\leftarrow}^{\mathrm{Su}}\{T_.^j\}.$$

For $\{S_.^i\} \in$ pro-(s. sets$_{*c}$) *weakly equivalent in* pro-$\mathcal{H}_*$ *to* $\{R_.^m\}$ *with* $H^k(R_.^m, Z/\ell)$ *finite for each* $R_.^m$ *and each* $k > 0$, *there is a natural isomorphism in* $\mathcal{H}_*$

$$\operatorname{holim}_{\leftarrow}\{(Z/\ell)_\infty(S_.^i)\} \to \operatorname{holim}_{\leftarrow}^{\mathrm{Su}}(\{S_.^i\}^{\hat{\ell}}).$$

Proof. If $T_. \in$ (s. sets$_{*c}$) has finite homotopy groups, then $T_. \to (Z/\ell)_\infty(T_.)$ induces an isomorphism in Z/ℓ cohomology and $\pi_k((Z/\ell)_\infty(T_.))$ is a finite ℓ-group for each $k > 0$ ([13], VII.4.3). Because any $Z_. \in \ell\mathcal{H}_{*c}$ is nilpotent, obstruction theory applied to such a $Z_.$ implies that $\operatorname{Hom}_{\mathcal{H}_*}((Z/\ell)_\infty(T_.), Z_.) \to \operatorname{Hom}_{\mathcal{H}_*}(T_., Z_.)$ is bijective. Thus, $T_. \to (Z/\ell)_\infty(T_.)$ represents $T_. \to (T_.)^{\hat{\ell}}$.

If $T_. \in \mathcal{H}_{*c}$ is such that $H^k(T_., Z/\ell)$ is finite for each $k > 0$, then $(Z/\ell)\circ\cdots\circ(Z/\ell)(T_.)$ has finite ℓ-torsion homotopy groups for any iteration of the functor $(Z/\ell)(\ )$ ([13], I.2.3), so that $(Z/\ell)_n(T_.) = \operatorname{Tot}_n((\underset{\sim}{Z/\ell})(T_.))$ has finite homotopy groups for each $n > 0$ ([13], X.6.3). Moreover, $T_. \to \{\operatorname{cosk}_{n+1}(Z/\ell)_n(T_.)\,;\ n > 0\}$ is left final among maps in $\mathcal{H}_{*c}$ from $T_.$ to simplicial sets with finitely many non-zero homotopy groups each of which is ℓ-torsion ([13], VIII.8.3). Thus, $T_. \to \{(Z/\ell)_n(T_.)\}$ induces $T_.^{\hat{\ell}} \to \{(Z/\ell)_n(T_.)\}$ which is a weak equivalence in pro-$\mathcal{H}_*$.

If $\{T_.^j\} \in$ pro-(s. sets$_{*c}$) is such that each $T_.^j$ is a Kan complex and $\pi_k(\{T_.^j\})$ is pro-finite, then the existence of the map $\operatorname{holim}_{\leftarrow}\{T_.^j\} \to \operatorname{holim}_{\leftarrow}^{\mathrm{Su}}\{T_.^j\}$ commuting in $\mathcal{H}_*$ with the projections to $T_.^j$ is given by the characterizing property of $\operatorname{holim}_{\leftarrow}^{\mathrm{Su}}(\ )$ and [13], XI.3.4. The fact that this map induces an isomorphism in homotopy groups follows from the spectral sequence of Theorem 6.8 because $\lim_{\leftarrow}^s\{\pi_k(T_.^j)\} = 0$ for $s,k > 0$. Because $\operatorname{holim}_{\leftarrow}\{T_.^j\}$ is connected by Proposition 6.9, we conclude that $\operatorname{holim}_{\leftarrow}\{T_.^j\} \to \operatorname{holim}_{\leftarrow}^{\mathrm{Su}}\{T_.^j\}$ is a homotopy equivalence.

Finally, the last asserted isomorphism in $\mathcal{H}_*$ follows from the previous isomorphism applied to $\{T_.^j\} = \{(Z/\ell)_n(S_.^i)\,;\ n > 0, i \in I\}$ plus the observation that the natural map $(\mathbf{Z}_\ell)_\infty(T.) \to \underleftarrow{\text{holim}}(\mathbf{Z}/\ell)_n(T.)$ is a homotopy equivalence for any $T. \in \mathcal{H}_{*c}$. ■

7. FINITENESS AND HOMOLOGY

In this chapter we consider *noetherian simplicial schemes* (i.e., simplicial schemes which are noetherian in each dimension). As we verify in Corollary 7.2, the etale topological type of such a noetherian simplicial scheme is weakly equivalent to a pro-object in the homotopy category of simplicial sets which are finite in each dimension. In Theorem 7.3, we generalize to simplicial schemes the criterion of M. Artin and B. Mazur that the homotopy pro-groups of the etale topological type be pro-finite. In Proposition 7.5, we show that pro-objects of finite abelian groups are anti-equivalent to torsion abelian groups. Under this anti-equivalence, the homology pro-groups with coefficients in the dual of a sheaf of constructible abelian groups (as defined in Definition 7.4) correspond to cohomology. A theorem of P. Deligne provides examples of simplicial schemes with finite homology groups with various coefficients.

PROPOSITION 7.1. *Let* $X.\ ,x$ *be a pointed noetherian simplicial scheme. Then the full subcategories*

$$nHR(X.) \subset HR(X.)\,, \quad nHR(X.\ ,x) \subset HR(X.\ ,x)$$

consisting of hypercoverings $U.. \to X.$ *with* $U_{s,t}$ *noetherian for all* $s,t \geq 0$ *("noetherian hypercoverings") are left filtering and the inclusion functors are left final.*

Proof. If $U \to X_n$ and $V \to X_n$ are etale with U and V noetherian, then these maps are of finite type; thus, the fibre product of $U \to X_n \leftarrow V$ is also of finite type over X_n and therefore is also noetherian. We conclude that $nHR(X.)$ and $nHR(X.\ ,x)$ have products. Consequently, it suffices to prove that the inclusion functors are left final in order to conclude that $nHR(X.)$ and $nHR(X.,x)$ are left filtering.

For any $U \to X_n$ in $Et(X.)$, U is noetherian if and only if U is a union of finitely many connected components. Therefore, it suffices to prove for any pointed hypercovering $U..,u \to X..,x$ the existence of $V..,v \to U..,u$ such that $V_{s,t} \to U_{s,t}$ is the inclusion of finitely many components of $U_{s,t}$ for all $s,t \geq 0$ and $V..,v \to X..,x$ is a pointed hypercovering. We define $V_{0,0} \to U_{0,0}$ to be the pointed inclusion of finitely many components of $U_{0,0}$ which cover X_0. Proceeding inductively, we define $V_{s,t} \to U_{s,t}$ to be the inclusion of those connected components of $U_{s,t}$ which are in the image of $V_{s-1,t}$ or $V_{s,t-1}$ under some degeneracy map of $U..$ together with finitely many connected components of the fibre product of $(\mathrm{cosk}_{t-1}V_{s.})_t \to (\mathrm{cosk}_{t-1}U_{s.})_t \leftarrow U_{s,t}$ whose images cover $(\mathrm{cosk}_{t-1}V_{s.})_t - (\mathrm{sk}_{t-1}V_{s.})_t$ for any $s,t \geq 0$ (where $\mathrm{cosk}_{-1}U_{s.} = X_s = \mathrm{cosk}_{-1}V_{s.}$). ■

As an immediate corollary of Proposition 7.1 and Corollary 6.3, we obtain the following finiteness property of the weak homotopy type of $(X.,x)_{et}$.

COROLLARY 7.2. *Let $X.,x$ be a pointed, connected, noetherian simplicial scheme. Define $(X.,x)_{nht}$ in pro-$\mathcal{H}_*$, the noetherian (etale) homotopy type of $X.,x$, to be*

$$(X.,x)_{nht} = \pi \circ \Delta : nHR(X.,x) \to (s.\ sets_*)\ .$$

Then $(X.,x)_{nt}$ is a pro-object in the homotopy category of pointed simplicial sets which are finite in each dimension. Furthermore, the natural maps

$$(X.,x)_{nht} \leftarrow (X.,x)_{ht} \to (X.,x)_{et}$$

are weak equivalences in pro-$\mathcal{H}_*$. ■

We remind the reader that a scheme X is said to be geometrically unibranched if the integral closure of each of the local rings of X (stalks of the structure sheaf in the Zariski topology) is also a local ring. In particular, if each of these local rings is already integrally closed (as is the case if they are regular local rings), then X is geometrically unibranched.

The following theorem is an easy generalization of a theorem of M. Artin and B. Mazur ([8], 11.2). We shall find this theorem particularly useful when we consider function complexes.

THEOREM 7.3. *Let* $X.\,, x$ *be a pointed, noetherian simplicial scheme such that* X_n *is connected and geometrically unibranched for each* $n \geq 0$. *Then any pointed, noetherian hypercovering* $U..\,, u \to X.\,, x$ *has the property that* $\pi_k(\pi(\Delta U..), u)$ *is finite for each* $k > 0$. *Consequently,* $\pi_k((X.\,, x)_{et})$ *is a pro-finite group (i.e., isomorphic in pro-(grps) to a pro-object in the category of finite groups).*

Proof. For any pointed, noetherian hypercovering $U..\,, u \to X.\,, x$ and any $n \geq 0$, $U_{n.} \to X_n$ is a pointed, noetherian hypercovering so that the Artin-Mazur theorem implies that $\pi_k(\pi(U_{n.}))$ is finite for each $k > 0$. (Because X_n is geometrically unibranched, any connected etale open $U \to X$ is irreducible. The starting point of their proof is the observation that this implies that $\pi(U_{n.}) = \pi(U_{n.} \times_{X_n} \operatorname{Spec} K)$, where K is the field of fractions at the generic point of X_n.) Because each $\pi(U_{n.})$ is connected by Proposition 5.2, we may apply the homotopy spectral sequence of [12], B.5, to conclude that the homotopy groups of $\pi_k(\pi(\Delta U..), u)$ are also finite. Corollary 7.2 now immediately implies that $\pi_k((X.\,, x)_{et})$ is pro-finite for each $k > 0$. ∎

The necessity of the hypothesis that each X_n be connected in Theorem 7.3 can be readily understood by examining the case in which $X. = \operatorname{Spec} K \otimes S^1$.

Our approach to the homology of an abelian copresheaf is motivated by Corollary 3.10. We shall find it sufficient to consider copresheaves which commute with finite disjoint unions because we shall restrict our attention to noetherian simplicial schemes; the reader should observe that the dual of a sheaf does not commute with arbitrary disjoint unions.

DEFINITION 7.4. Let X. be a simplicial scheme. An *abelian copresheaf* on Et(X.) is a functor $P: Et(X.) \to$ (ab. grps) with the property that $P(U \amalg V) = P(U) \oplus P(V)$ for any $U \to X_n$, $V \to X_n$ in Et(X.). For any $i \geq 0$, the i-th *homology pro-group* of X. with values in the abelian copresheaf P is the pro-group

$$H_i(X.\ ,P) = H_i \circ P(\ ) : HR(X.) \to \text{(ab. grps)}$$

sending $U.. \to X.$ to the i-th homology group of the bicomplex P(U..). ∎

We shall be especially interested in the following two examples of abelian copresheaves. If M is a locally constant abelian sheaf on Et(X.), we define

$$M^0 : Et(X.) \to \text{(ab. grps)}$$

by setting $M^0(U) = \oplus M^0(U_\alpha)$, where the direct sum is indexed by the connected components U_α of U and where $M^0(U_\alpha) = M(U_\alpha)$ whenever M restricted to U_α is constant and $M^0(U_\alpha) = 0$ otherwise; we define $M^0(U) \to M^0(V)$ associated to a map $U \to V$ in Et(X.) to be the map whose restriction to $M^0(U_\alpha)$ is given by the inverse of $M(V_\beta) \to M(U_\alpha)$, where V_β is the component of V containing the image of U_α if M restricted to U_α is constant.

If F is an abelian presheaf on Et(X.) such that $F(U \amalg V) = F(U) \times F(V)$, then we define the *dual copresheaf*

$$F^\vee : Et(X.) \to \text{(ab. grps)}$$

by $F^\vee(U) = \text{Hom}(F(U), \mathbf{Q}/\mathbf{Z})$.

If $P: Et(X.) \to$ (ab. grps) takes values in the category of finite abelian groups (denoted (f. ab. grps)), then the left finality of $nHR(X.) \to HR(X.)$ implies that $H_i(X.\ ,P)$ is pro-finite for each $i \geq 0$. Such pro-finite abelian groups are not so unfamiliar, as we show in the next proposition.

PROPOSITION 7.5. *Let* $(\ )^\vee = \mathrm{Hom}(\ ,\mathbf{Q}/\mathbf{Z})$. *Then the functor*

$$\mathrm{colim} \circ (\ )^\vee : (\text{pro-(f. ab. grps)})^0 \to (\text{tor. ab. grps})$$

is an equivalence of categories from the opposite category of the category of pro-objects of finite abelian groups ("pro-finite abelian groups") to the category of torsion abelian groups. In particular, $\{A_i\} \in$ pro-(f. ab. grps) *is finite (i.e., isomorphic to a finite group) if and only if* $\mathrm{colim}\{A_i^\vee\}$ *is finite if and only if* $\lim\limits_{\leftarrow}\{A_i\}$ *is finite.*

Proof. Because any torsion abelian group A is the colimit of its finite subgroups, because the category of finite subgroups of A is left directed (where B maps to C in this category if and only if B contains C), and because $(\ )^\vee$ is a (contravariant) involution of (f. ab. grps), the functor $\mathrm{colim} \circ (\ )^\vee$ is essentially surjective.

To prove $\mathrm{colim} \circ (\ )^\vee$ is faithful, consider $f,g : \{G_i;\ i \in I\} \to \{H_j;\ j \in J\}$ in pro-(f. ab. grps). Then $f = g$ if and only if $f_j = g_j$ in $\mathrm{colim}\,\mathrm{Hom}(G_i,H_j)$ for each $j \in J$ if and only if $f_j^\vee = g_j^\vee$ in $\mathrm{colim}\,\mathrm{Hom}(H_j^\vee,G_i^\vee)$ for each $j \in J$ if and only if $f_j^\vee = g_j^\vee$ in $\mathrm{Hom}(H_j^\vee, \mathrm{colim}\{G_i^\vee\})$ for each $j \in J$ (because $H_j^\vee$ is finite) if and only if $\mathrm{colim}\, f_j^\vee = \mathrm{colim}\, g_j^\vee$ in $\mathrm{Hom}(\mathrm{colim}\{H_j^\vee\},\mathrm{colim}\{G_i^\vee\})$.

To prove $\mathrm{colim} \circ (\ )^\vee$ is fully faithful, consider $f : \mathrm{colim}\{H_j^\vee\} \to \mathrm{colim}\{G_i^\vee\}$ and write $f = \lim\limits_{\leftarrow} f_j$ with $f_j : H_j^\vee \to \mathrm{colim}\{G_i^\vee\}$. Since $H_j^\vee$ is finite, f_j factors through the image of some $G_m^\vee \to \mathrm{colim}\{G_i^\vee\}$; and since $G_m^\vee$ is finite, this image factors through some $G_k^\vee \to \mathrm{colim}\{G_i^\vee\}$. Thus, f_j factors through some map $f_{j,k} : H_j^\vee \to G_k^\vee$. If we let $f_j^\vee : \{G_i\} \to H_j$ be the map determined by $f_{j,k}^\vee$, then the fact that $\mathrm{colim} \circ (\ )^\vee$ is faithful implies that $\{f_j^\vee\} : \{G_i\} \to \{H_j\}$ is a well-defined map and that $\mathrm{colim} \circ (\ )^\vee(\{f_j^\vee\}) = f$.

In particular, the fact that $\operatorname{colim} \circ (\ )^\vee$ is an equivalence implies that $\operatorname{colim}\{A_i^\vee\} = A$ is finite if and only if $\{A_i\}$ is isomorphic to a finite group A. Since $(\operatorname{colim}\{A_i^\vee\})^\vee = \lim\limits_{\leftarrow}\{A_i\}$, $\operatorname{colim}\{A_i^\vee\}$ is finite if and only if $\lim\limits_{\leftarrow}\{A_i\}$ is finite. ∎

We next utilize the equivalence of Proposition 7.5 to relate homology to cohomology. We remind the reader that a sheaf F on a noetherian scheme X (with the etale topology) is said to be *constructible* if there exists a finite collection of locally closed subschemes $\{X_i\}$ of X whose disjoint union is X such that F restricted to each X_i is locally constant with finite stalks. More generally, if $X.$ is a noetherian simplicial scheme and F is a sheaf on $Et(X.)$, then F is said to be constructible if F^n is constructible on $Et(X_n)$ for each $n \geq 0$. If F is constructible, then $F(U)$ is a finite set for any $U \to X_n$ in $Et(X.)$ with U noetherian.

PROPOSITION 7.6. *Let* $X.$ *be a noetherian simplicial scheme, and let* F *be a constructible abelian sheaf on* $Et(X.)$. *For each* $i \geq 0$, *there is a natural duality isomorphism*

$$(\operatorname{colim} \circ (\ )^\vee)(H_i(X.\,,F^\vee)) \simeq H^i(X.\,,F)\,,$$

so that $H_i(X.\,,F^\vee)$ *is finite if and only if* $H^i(X.\,,F)$ *is finite. Moreover, if* M *is a locally constant constructible abelian sheaf on* $Et(X.)$, *then for all* $i \geq 0$ *there are natural isomorphisms*

$$(\operatorname{colim} \circ (\ )^\vee)(H_i(X.\,,M^0)) \simeq H^i(X.\,,M)\,.$$

Proof. If $U.. \to X.$ is a noetherian hypercovering and F a constructible sheaf, then $(F^\vee(U..))^\vee$ is isomorphic to $F(U..)$ so that $(H_i(F^\vee(U..)))^\vee \simeq H^i(F(U..))$ for any $i \geq 0$. Consequently,

$$\operatorname{colim}(H_i(F^\vee(U..))^\vee) \simeq \operatorname{colim} H^i(F(U..)) \simeq H^i(X.\,,F)$$

where the colimits are indexed by $nHR(X.)$.

If M is a locally constant, constructible abelian sheaf on $Et(X.)$, then $M^0(U..)$ is naturally isomorphic to $M^\vee(U..)$ whenever $U..$ is a noetherian hypercovering such that M restricted to $U_{0,0}$ is constant. Consequently,

$$\operatorname{colim}(H_i(M^0(U..))^\vee) \simeq \operatorname{colim}(H_i(M^\vee(U..))^\vee) \simeq H^i(X. ,M) . \blacksquare$$

Using Proposition 7.6, we immediately conclude the homology analogue of the spectral sequence of Proposition 2.4

$$E^1_{s,t} = H_t(X_s, F^{\vee s}) \Rightarrow H_{s+t}(X, F^\vee)$$

whenever $X.$ is noetherian and F is constructible. We can also conclude, for example, the homology analogue of Proposition 3.7 asserting that

$$H_*(X. ,F^\vee) \simeq H_*(\Delta U.. ,F^\vee)$$

whenever $U.. \to X.$ is a noetherian hypercovering and F is constructible.

We conclude this chapter with the following theorem whose proof is an immediate consequence of P. Deligne's criterion for the finiteness of cohomology ([24], "Finitude" 1.1) (extended to simplicial schemes using Proposition 2.4) and Proposition 7.6.

THEOREM 7.7. *Let* $X.$ *be a simplicial scheme of finite type over* $\operatorname{Spec} R$, *where* R *is a complete discrete valuation ring with algebraically closed residue field. Then for all* $i \geq 0$, $H^i(X. ,F)$ *and* $H_i(X. ,F^\vee)$ *are finite for any constructible abelian sheaf* F *on* $Et(X.)$ *with stalks of order invertible in* R. $\blacksquare$

The condition on the orders of the stalks of F in Theorem 7.7 is necessary. For example, $H^1(\operatorname{Spec} k[x], \mathbf{Z}/p)$ is infinite whenever k is an infinite field of characteristic p.

8. COMPARISON OF HOMOTOPY TYPES

In Chapter 5, we identified various homotopy invariants of the etale topological type $(X.\ ,x)_{et}$ in terms of algebraic invariants. In Chapter 6, we described the homotopy-theoretic context in which these invariants play a central role. We now proceed to employ this material to translate various theorems concerning etale cohomology groups and fundamental groups into theorems concerning the homotopy type of $(X.\ ,x)_{et}$.

We begin by comparing in Proposition 8.1 the homotopy type of $(X.\ ,x)_{et}$ to that of $(\Delta U..\ ,u)_{et}$, where $U..\ ,u \to X.\ ,x$ is a pointed hypercovering. The proof of Proposition 8.1 is representative of the method of proof of each of the results in this chapter. For a simplicial scheme $X.$ over the complex numbers, Theorem 8.4 presents the very useful comparison of the homotopy type of $(X.\ ,x)_{et}$ with the homotopy type of its underlying simplicial space with the "classical topology." Propositions 8.6 and 8.7 obtain homotopy theoretic conclusions from the proper base change theorem and smooth base change theorem for etale cohomology. Finally, in Proposition 8.8, we investigate reductive group schemes and their classifying spaces.

In this first proposition, we conclude that from a homotopy-theoretic point of view a simplicial scheme may be replaced by one of its hypercoverings. The usefulness of this conclusion lies in the fact that the hypercovering may consist of schemes which are simpler (being more local) than the original simplicial scheme. For example, this is the basic observation underlying the discussion of tubular neighborhoods of Chapter 15. A somewhat weakened form of the following proposition was proved by D. Cox in [18], IV.2.

PROPOSITION 8.1. *Let* $X.\,,x$ *be a pointed, connected, locally noetherian simplicial scheme, and let* $g : U..\,,u \to X.\,,x$ *be a pointed hypercovering. Then*

$$g_{et} : (\Delta U..\,,u)_{et} \to (X.\,,x)_{et}$$

is a strict map of pro-(s. sets$_{*c}$) *which is a weak equivalence in* pro-$\mathcal{H}_*$. *In particular, for all* $i > 0$, $g_* : \pi_i((\Delta U..\,,u)_{et}) \to \pi_i((X.\,,x)_{et})$ *is an isomorphism, and if these pro-groups are pro-finite, then*

$$(\operatorname{holim}_{\leftarrow} \circ \operatorname{Sing.}(\ ) \circ |\ |)(g) : \operatorname{holim}_{\leftarrow}(\operatorname{Sing.}(|(\Delta U..\,,u)_{et}|)) \to \operatorname{holim}_{\leftarrow}(\operatorname{Sing.}(|(X.\,,x)_{et}|))$$

is a homotopy equivalence (where $|\ | :$ (s. sets) $\to$ (top. spaces) *is the geometric realization functor and* Sing.( ) : (top. spaces) $\to$ (s. sets) *is the singular functor as discussed in* [42]).

Proof. By Propositions 3.7 and 5.9, $g : \Delta U.. \to X.$ induces an isomorphism $H^*((X.)_{et},M) \xrightarrow{\sim} H^*((\Delta U..)_{et},M)$ for any abelian local coefficient system M on $(X.)_{et}$. By Theorem 6.3, to prove that g_{et} is a weak equivalence in pro-$\mathcal{H}_*$, it suffices to prove that g also induces an isomorphism $\pi_1((\Delta U..\,,u)_{et}) \to \pi_1((X.\,,x)_{et})$. By Proposition 5.6, it suffices to prove that g induces an equivalence of categories $\Pi(X.\,,G) \xrightarrow{\sim} \Pi(\Delta U..\,,G)$ for any group G. As recalled in the proof of Proposition 5.6, there is an equivalence of categories $\Pi(X_s,G) \xrightarrow{\sim} \Pi(U_{s.}\,,G)$ for any $s \geq 0$. Consequently, the required equivalence follows from Lemma 5.4 and Proposition 5.5, which provide the intermediate equivalences $\Pi(X.\,,G) \xrightarrow{\sim}$ $\langle \Pi(X_0,G);\ \text{d. data} \rangle$ and $\langle \Pi(U_{0.}\,,G);\ \text{d. data} \rangle \simeq \Pi(\Delta U..\,,G)$.

Theorem 6.2 now implies that g_{et} induces an isomorphism on homotopy pro-groups, whereas Proposition 6.9 implies that $(\operatorname{holim}_{\leftarrow} \circ \operatorname{Sing.} \circ |\ |)(g)$ is a homotopy equivalence (because the natural map $S. \to \operatorname{Sing.}(|S.|)$ is a weak equivalence with $\operatorname{Sing.}(|S.|)$ a Kan complex for any simplicial set $S.$). ∎

Most applications of etale homotopy theory have utilized the "Čech topological type," $\Delta(X.\,,x)_{ret}$, as defined preceding Definition 4.4. We verify that $\Delta(X.\ ,x)_{ret}$ has the same weak homotopy type as $(X.\ ,x)_{et}$ for "most" pointed simplicial schemes $X.\ ,x$.

PROPOSITION 8.2. *Let* $X.\ ,x$ *be a pointed, connected, locally noetherian simplicial scheme, and let* $\Delta\circ(X.\ ,x)_{ret}$ *denote the Čech topological type (as defined above). Then there is a natural strict map of* pro-(s. sets$_{*c}$)

$$(X.\ ,x)_{et} \to \Delta\circ(X.\ ,x)_{ret}$$

which induces an isomorphism $\pi_1((X.\ ,x)_{et}) \xrightarrow{\sim} \pi_1(\Delta\circ(X.\ ,x)_{ret})$. *Furthermore, if* X_n *is quasi-projective over a noetherian ring for each* $n \geq 0$, *then this map is a weak equivalence in* pro-$\mathcal{H}_*$.

Proof. The asserted strict map is induced by the rigid analogue of $\cosk_0^{X.}(\ ):C(X.) \to HR(X.)$. The fact that this map induces an isomorphism on fundamental pro-groups follows directly from Corollary 5.7. Arguing as in the proof of Proposition 5.9, we easily conclude the natural isomorphism $\check{H}^*(X.\ ,M) \xrightarrow{\sim} H^*(\Delta\circ(X.\ ,x)_{ret},M)$ for any locally constant, abelian sheaf M on $Et(X.)$. Thus, if X_n is quasi-projective over a noetherian ring for each $n \geq 0$, then Theorem 3.9 implies that $(X.\ ,x)_{et} \to \Delta\circ(X.\ ,x)_{ret}$ induces an isomorphism in cohomology with any abelian local coefficients. Consequently, for a pointed simplicial scheme, the fact that $(X.\ ,x)_{et} \to \Delta\circ(X.\ ,x)_{ret}$ is a weak equivalence in pro-$\mathcal{H}_*$ follows from Theorem 6.2. ∎

We next proceed to consider simplicial schemes $X.$ of finite type over $\mathbb{C}$ (i.e., X_n is of finite type over Spec $\mathbb{C}$, where $\mathbb{C}$ denotes the complex numbers). If X is of finite type over $\mathbb{C}$, then X^{top} is the set of complex points of X with the usual (analytic) topology; if $X.$ is of finite type over $\mathbb{C}$, then $X.^{top}$ is the *simplicial space* (i.e., the simplicial object of topological spaces) with $(X.^{top})_n = X_n^{top}$.

We recall that the *geometric realization* of a simplicial space $T.$, $|T.|$, is the quotient of the topological space $\coprod_{n\geq 0} T_n \times \Delta[n]$ by the equivalence relation $(t, \alpha(x)) \sim (\alpha(t), x)$ for any $\alpha : \Delta[n] \to \Delta[m]$ in Δ (where $T_n \times \Delta[n]$ is given the product topology and $\Delta[n] = \{x = (x_0, \cdots, x_n) : \Sigma x_i = 1, x_i \geq 0\} \subset R^{n+1}$). We also recall that if $S..$ is a bi-simplicial set and if $\{n \mapsto |S_{n.}|\}$ is the associated simplicial space, then $|\Delta(S..)|$ (the geometric realization of the diagonal simplicial set) is homeomorphic to $|\{n \mapsto |S_{n.}|\}|$ (cf. [65], 1).

We now give a technical lemma relating the singular cohomology of the geometric realization of a simplicial space $T.$ to its sheaf cohomology. In analogy with Definition 1.4, we define the *local homeomorphism site* $Lh(T.)$ as follows. As a category, $Lh(T.)$ has objects which are local homeomorphisms $W \to T_n$ for some $n \geq 0$; a map in $Lh(T.)$ is a commutative square

$$\begin{array}{ccc} W & \longrightarrow & Z \\ \downarrow & & \downarrow \\ T_n & \longrightarrow & T_m \end{array}$$

with $T_n \to T_m$ a specified structure map of $T.$; a covering of $W \to T_n$ is defined to be a family of local homeomorphisms $\{W_i \to W\}$ over T_n whose images of W_i in W cover W. As in Definition 2.3, for any $i \geq 0$ we define

$$H^i_{\ell h}(T. \, , \quad) : AbSh(T.) \to Ab$$

to be the i-th right derived functor of the functor sending an abelian sheaf F on $Lh(T.)$ to the kernel of the map $d_0^* - d_1^* : F(T_0) \to F(T_1)$.

LEMMA 8.3. *Let* $T.$ *be a simplicial space with* T_n *paracompact for each* $n \geq 0$. *For any locally constant abelian sheaf* M *on* $Lh(T.)$, *there are natural isomorphisms*

$$H^*_{\ell h}(T. ,M) \simeq H^*(\Delta(\mathrm{Sing.}(T.)),M) \simeq H^*(\mathrm{Sing.}(|T.|),M)$$

where $\mathrm{Sing.}(T.)$ *is the bisimplicial set given in bidegree* s,t *by* $\mathrm{Sing}_t(T_s)$.

Proof. Because every local homeomorphism $W \to T_n$ is covered by a family $\{W_i \to W\}$ with each $W_i \to W \to T_n$ an open immersion, the categories of abelian sheaves on Lh(T.) and on the analogous site $\mathcal{O}i(T.)$ of open immersions are equivalent. We define a sheaf $\mathcal{S}^p M$ on $\mathcal{O}i(T.)$ for any $p \geq 0$ as the sheaf associated to the presheaf sending an open immersion $W \to T_n$ to $C^p(\text{Sing.}(W),M)$ ([14], I.7). Then $M \to \mathcal{S}^{\cdot} M$ is resolution with the property that $(\mathcal{S}^p M)_n$ is acyclic on $\mathcal{O}i(T_n)$ for any $p,n \geq 0$ ([14], III.1). Using the local homeomorphism analogue of Proposition 2.4, we conclude that $H^*_{\ell h}(T.\,,M)$ is naturally isomorphic to $H^*(\mathcal{S}^{\cdot} M(T.))$.

On the other hand, the natural map $C^*(\text{Sing.}(T_n),M) \to \mathcal{S}^{\cdot} M(T_n)$ induces an isomorphism in cohomology ([14], 1.7) for each $n \geq 0$, so that the natural map of bicomplexes $C^*(\text{Sing.}(T.),M) \to \mathcal{S}^{\cdot} M(T.)$ induces an isomorphism in cohomology $H^*(\Delta(\text{Sing.}(T.)),M) \to H^*_{\ell h}(T.\,,M)$. Thus, to complete the proof, it suffices to verify that the natural map $\Delta(\text{Sing.}(T.)) \to \text{Sing.}(|T.|)$ induces an isomorphism in cohomology. This is readily shown by comparing the cohomology of $|\Delta(\text{Sing.}(T.))|$ with that of the geometric realization of the simplicial space $\{n \mapsto |\text{Sing.}(T_n)|\}$ with that of $|T.|$. ∎

The main ingredient of the proof of the following theorem is the "classical comparison theorem" for etale cohomology proved by M. Artin and A. Grothendieck and the "Riemann existence theorem" proved by H. Grauert and R. Remmert ([7], XI.4.3). Various (weaker) versions of Theorem 8.4 have been proved first by M. Artin and B. Mazur ([8], 12.9), then by the author, R. Hoobler and D. Rector, and D. Cox.

THEOREM 8.4 (Comparison Theorem). *Let* $X.\,,x$ *be a pointed, connected simplicial scheme of finite type over* $\mathbb{C}$, *and let* $X.^{top},x$ *denote the associated simplicial space (as above). There exist strict maps of* pro-(s. sets$_{*c}$)

$$(X.\,,x)_{et} \xleftarrow{\rho} (X.\,,x)_{s.et} \xrightarrow{\tau} \text{Sing.}(|X.^{top},x|)$$

such that τ *is an isomorphism in* pro-$\mathcal{H}_*$, *and* $(\rho)^{\hat{P}}$ *is a weak equivalence in* pro-$\mathcal{H}_*$, *where* P *is the set of all primes (cf. Corollary 6.5).*

Proof. We define $(X_.\,,x)_{s.et}$ in pro-(s. sets_*) to be the functor

$$\Delta \circ \text{Sing.}(\ )\circ(\ )^{top}\circ\Delta : HRR(X.) \to (\text{s. sets}_*)$$

sending $U.. \to X.$ to the diagonal of the bisimplicial set $\text{Sing.}(\Delta U_{..}^{top})$. The map τ is the composition of the natural map $\eta: \Delta(\text{Sing.}(\Delta U_{..}^{top})) \to \Delta(\text{Sing.}(X_.^{top}))$ (for any $U.. \to X.$) and the canonical homotopy equivalence $\Delta(\text{Sing.}(X.)) \to \text{Sing.}(|X_.^{top}|)$ (discussed in the proof of Lemma 8.3). Using descent for principal homogeneous spaces over X_s^{top}, the argument of [8], 10, applies to show that $\Pi(X_s^{top},G)$ is equivalent to $\Pi(U_{s.}^{top},G)$. As argued in the proof of Proposition 8.1, this implies that $\Pi(\Delta U_{..}^{top},G)$ is equivalent to $\Pi(X_.^{top},G)$, which implies that $\pi_1(|\Delta U_{..}^{top},u|) \simeq \pi_1(\Delta(\text{Sing.}(U_{..}^{top},u)))$ is isomorphic to $\pi_1(\Delta(\text{Sing.}(X_.^{top},x))) \simeq \pi_1(|X_.^{top},x|)$. Consequently, η induces an isomorphism of fundamental groups.

As argued in Corollary 5.8, an abelian local coefficient system M on $\Delta(\text{Sing.}(X_.^{top}))$ is in one-to-one correspondence with a locally constant abelian sheaf on $Lh(X_.^{top})$. By Lemma 8.3 and the local homeomorphism analogue of Proposition 3.7, η also induces an isomorphism in cohomology with abelian local coefficients. Therefore, η is a weak homotopy equivalence for any $U.. \to X.$ in HRR(X.), so that τ is an isomorphism in pro-$\mathcal{H}_*$.

The map ρ is induced by the natural transformation

$$\Delta\circ\text{Sing.}(\ )\circ(\ )^{top}\circ\Delta \to \pi\circ\Delta$$

defined by sending $\alpha: \Delta[k] \to U_{k,k}^{top}$ in $(\Delta(\text{Sing.}(\Delta U_{..}^{top})))_k$ to the connected component in $(\pi(\Delta U..))_k$ containing the image of α. To prove that $(\rho)^{\hat{P}}$ is a weak equivalence in pro-$\mathcal{H}_*$, we factor ρ as the composition

$$\delta\circ\gamma\circ\beta^{-1}: (X_.\,,x)_{s.et} \to (X_.\,,x)_{s.\ell h} \to (X_.\,,x)_{\ell h} \to (X_.\,,x)_{et}$$

defined as follows. Let $HRR(X_.^{top})$ be defined as in Proposition 4.3 with the site Et(X.) replaced by $Lh(X_.^{top})$. Define $(X_.\,,x)_{s.\ell h}$ to be

$\Delta \circ \mathrm{Sing.}(\ )\circ\Delta : \mathrm{HRR}(X_.^{top}) \to (\text{s. sets}_*)$ and define $(X.\ ,x)_{\ell h}$ to be $\pi\circ\Delta : \mathrm{HRR}(X_.^{top}) \to (\text{s. sets}_*)$. Define $\beta : (X.\ ,x)_{s.\ell h} \to (X.\ ,x)_{s.et}$ to be the isomorphism in pro-$\mathcal{H}_*$ (as argued above, both are isomorphic to $\mathrm{Sing.}(|X_.^{top},x|)$) determined by the forgetful functor $\mathrm{HRR}(X.) \to \mathrm{HRR}(X_.^{top})$. Define γ exactly as we defined ρ, and define δ to be also induced by the forgetful functor. Then, $\rho\circ\beta = \delta\circ\gamma : (X.\ ,x)_{s.\ell h} \to (X.\ ,x)_{et}$.

As in Proposition 5.6, we identify $\pi_1((X.\ ,x)_{\ell h})$ with $\pi_1(|X_.^{top},x|)$, so that γ induces an isomorphism on fundamental pro-groups. As in Proposition 5.9, we identify $H^*_{\ell h}(X_.^{top},M)$ with $H^*((X.\ ,x)_{\ell h},M)$ for any locally constant abelian sheaf on $\mathrm{Lh}(X_.^{top})$, so that Lemma 8.3 implies that γ induces an isomorphism in cohomology with abelian local coefficients. Thus, γ is an isomorphism in pro-$\mathcal{H}_*$.

Finally, the Riemann existence theorem and the usual descent argument relating $\Pi(X_.^{top},G)$ to $<\Pi(X_0^{top},G)$; d. data$>$ and $\Pi(X.\ ,G)$ to $<\Pi(X_0,G)$; d. data$>$ imply that δ induces an isomorphism on the pro-P completions of the fundamental pro-groups. By the classical comparison theorem and Proposition 2.4, we conclude that δ induces an isomorphism in cohomology with abelian local coefficients whose stalks are finite. Thus, $(\delta)^{\hat{P}}$ is a weak equivalence in pro-$\mathcal{H}_*$. ■

COROLLARY 8.5. *Let* $X.\ ,x$ *be a pointed, connected simplicial scheme of finite type over* $\mathbb{C}$, *and let* ℓ *be a prime. Then* ρ *and* τ *of Theorem 8.4 induce homotopy equivalences*

$$\underset{\leftarrow}{\mathrm{holim}}\circ(Z/\ell)_\infty((X.\ ,x)_{et}) \leftarrow \underset{\leftarrow}{\mathrm{holim}}\circ(Z/\ell)_\infty((X.\ ,x)_{s.et}) \to (Z/\ell)_\infty(\mathrm{Sing.}(|X_.^{top},x|)).$$

Proof. By Proposition 6.10 and Corollary 7.2, it suffices to prove that ρ and τ induce homotopy equivalences

$$\underset{\leftarrow}{\mathrm{holim}}^{Su}((X.\ ,x)^{\hat{\ell}}_{et}) \leftarrow \underset{\leftarrow}{\mathrm{holim}}^{Su}((X.\ ,x)^{\hat{\ell}}_{s.et}) \to \underset{\leftarrow}{\mathrm{holim}}^{Su}(\mathrm{Sing.}\,|X_.^{top},x|^{\hat{\ell}}).$$

This follows immediately from Theorem 8.4 and Corollary 6.7. ■

We next provide the homotopy theoretic version of the proper base change theorem in etale cohomology ([7], XII.5.1). We recall that a *strict hensel local ring* is a local ring with a separably closed residue field which satisfies Hensel's lemma ([59], I.4). Such a local ring R has the property that $(\operatorname{Spec} R)_{et}$ is contractible.

PROPOSITION 8.6. *Let* R *be a strict hensel local ring, and let* X. *be a connected simplicial scheme over* Spec R *such that* $X_n \to \operatorname{Spec} R$ *is proper for each* $n \geq 0$. *Let* K *be a separably closed field containing the residue field of* R, *and let* $i: Y. \to X.$ *be defined by* $Y_n = X_n \underset{\operatorname{Spec} R}{\times} \operatorname{Spec} K$.

For any geometric point y *of* Y_0, *the strict map of* pro-(s. sets_{*c})

$$i : (Y. ,y)_{et} \to (X. ,y)_{et}$$

is such that $(i)^{\hat{P}}$ *is a weak equivalence in* pro-$\mathcal{H}_*$, *where* P *is the set of all primes. Consequently, for any prime* ℓ, i *induces a homotopy equivalence*

$$i : \underset{\leftarrow}{\text{holim}} \circ (Z/\ell)_\infty((Y. ,y)_{et}) \to \underset{\leftarrow}{\text{holim}} \circ (Z/\ell)_\infty((X. ,y)_{et}) .$$

Proof. The proper base change theorem implies that i_s induces an equivalence of categories $\Pi(Y_s,G) \xrightarrow{\sim} \Pi(X_s,G)$ for any finite group G. By Lemma 5.4 and Proposition 5.6, this implies that i induces a bijection $\operatorname{Hom}(\pi_1(X. ,x),G) \to \operatorname{Hom}(\pi_1(Y. ,y),G)$ for any finite group G. Moreover, the proper base change theorem and Proposition 2.4 imply that i induces an isomorphism $H^*(X. ,M) \xrightarrow{\sim} H^*(Y. ,M)$ for any locally constant, constructible abelian sheaf M on Et(X.). In particular, Y. is connected by Proposition 5.2. By Proposition 5.9, these cohomology isomorphisms imply that i induces an isomorphism $H^*((X.)_{et},M) \to H^*((Y.)_{et},M)$ for any abelian local coefficient system M on $(X.)_{et}$ with finite fibres. By Corollary 6.5, this implies that $(i)^{\hat{P}}$ is a weak equivalence in pro-$\mathcal{H}_*$. Consequently, Proposition 6.10 and Corollary 7.2 imply that $\underset{\leftarrow}{\text{holim}} \circ (Z/\ell)_\infty((Y. ,y)_{et}) \to \underset{\leftarrow}{\text{holim}} \circ (Z/\ell)_\infty((X. ,y)_{et})$ is a homotopy equivalence. ∎

The proof of Proposition 8.6 applies as well to prove the following theorem provided we employ the proper, smooth base change theorem ([7], XVI.2.2) rather than the proper base change theorem.

PROPOSITION 8.7. *Let* R *be a strict hensel local domain, and let* X. *be a connected simplicial scheme over* R *such that* $X_n \to \mathrm{Spec}\,R$ *is proper and smooth for each* $n \geq 0$. *Let* $\varepsilon : \mathrm{Spec}\,R \to X_0$ *be a section of the structure map* $X_0 \to \mathrm{Spec}\,R$, *let* $x : \mathrm{Spec}\,\Omega_x \to \mathrm{Spec}\,R \to X_0$ *be a geometric point over the closed point of* Spec R, *and let* $z : \Omega_z \to \mathrm{Spec}\,R \to X_0$ *be a geometric point over the generic point of* Spec R *(so that any etale neighborhood* $U \to X_0$ *of* x *is also an etale neighborhood of* z *). Let* $j : Z. \to X.$ *be defined by* $Z_n = X_n \underset{\mathrm{Spec}\,R}{\times} \mathrm{Spec}\,F$, *where* F *is any separably closed field containing* R. *Then* j *induces strict maps in* pro-(s. sets) *and* pro-$\mathcal{H}_{*c}$

$$j : (Z.)_{et} \to (X.)_{et}, \quad j : (Z.\,,z)_{ht} \to (X.\,,x)_{ht}$$

such that $(j)^{\hat{L}}$ *is a weak equivalence in* pro-$\mathcal{H}_*$, *where* L *is the set of all primes except the residue characteristic of* R. *Consequently, for any* $\ell \in L$, j *induces a homotopy equivalence*

$$j : \underleftarrow{\mathrm{holim}} \circ (\mathbb{Z}/\ell)_\infty((Z.)_{et}) \to \underleftarrow{\mathrm{holim}} \circ (\mathbb{Z}/\ell)_\infty((X.)_{et}) . \blacksquare$$

We conclude this chapter with the following specific comparison theorem which has proved useful in many applications (see, for example, the discussion of Chapter 9). The reader can consult [40] for an explicit discussion of $G_{\mathbb{Z}}$ and its cohomological properties.

PROPOSITION 8.8. *Let* $G(\mathbb{C})$ *be a complex reductive Lie group, and let* $G_{\mathbb{Z}}$ *be an associated Chevalley integral group scheme. Let* F *denote the algebraic closure of the prime field* F_p, *let* R *be the Witt vectors of* F *(a complete discrete valuation ring with residue field* F *), and let* $R \to \mathbb{C}$ *be a chosen embedding. Then for any algebraically closed field* k *containing* F *the base change maps* $BG_k \to BG_F \to BG_R \leftarrow BG_{\mathbb{C}}$ *determine strict maps in* pro-(s. sets$_{*c}$)

(8.8.1) $(BG_k)_{et} \to (BG_F)_{et} \to (BG_R)_{et} \leftarrow (BG_{\mathbb{C}})_{et} \leftarrow (BG_{\mathbb{C}})_{s.et} \to \mathrm{Sing.}(BG(\mathbb{C}))$

whose pro-L *completions are weak equivalences in* pro-$\mathcal{H}_*$, *where* $L = P - \{p\}$. *Consequently, for any prime* $\ell \neq p$, *these maps determine a chain of homotopy equivalences between*

$$\underleftarrow{\mathrm{holim}} \circ (\mathbb{Z}/\ell)_\infty((BG_k)_{et}) \quad \textit{and} \quad (\mathbb{Z}/\ell)_\infty \circ \mathrm{Sing.}(BG(\mathbb{C})) .$$

Proof. The fact that the base change maps $G_k \to G_F \to G_R \leftarrow G_{\mathbb{C}}$ induce isomorphisms in $\mathbb{Z}/m$ cohomology for $(m,p) = 1$ is proved in [40]. By Proposition 2.4, this implies that the maps $BG_k \to BG_F \to BG_R \leftarrow BG_{\mathbb{C}}$ also induce isomorphisms in $\mathbb{Z}/m$ cohomology for $(m,p) = 1$. Let $\bar{e}$ and e be geometric points of $(BG_R)_0 = \mathrm{Spec}\, R$ over the closed point and the geometric point respectively. These geometric points provide base points for BG_k, BG_F, BG_R, and $BG_{\mathbb{C}}$ in such a way that the chain of maps (8.8.1) consists of strict maps in pro-(s. sets_c) and pro-$\mathcal{H}_{*c}$. Because $(BG_k)_0$, $(BG_F)_0$, $(BG_R)_0$, and $(BG_{\mathbb{C}})_0$ are simply connected, and $(BG_k)_1$, $(BG_F)_1$, $(BG_R)_1$, and $(BG_{\mathbb{C}})_1$ are connected, Lemma 5.4 implies that $(BG_k)_{et}$, $(BG_F)_{et}$, $(BG_R)_{et}$, and $(BG_{\mathbb{C}})_{et}$ are simply connected. Thus, $(BG_k)_{et} \to (BG_F)_{et} \to (BG_R)_{et} \leftarrow (BG_{\mathbb{C}})_{et}$ have pro-L completions which are weak equivalences in pro-$\mathcal{H}_*$. The fact that $(BG_{\mathbb{C}})_{et} \leftarrow (BG_{\mathbb{C}})_{s.et} \to \mathrm{Sing.}(BG(\mathbb{C}))$ have pro-L completions which are weak equivalences is given by Theorem 8.4 (and the homotopy equivalence $\mathrm{Sing.}(|BG_{\mathbb{C}}^{top}|) \to \mathrm{Sing.}(BG(\mathbb{C}))$). Proposition 6.10 and Corollary 7.2 now imply that the chain of maps (8.8.1) determines a chain of homotopy equivalences relating $\underleftarrow{\mathrm{holim}} \circ (\mathbb{Z}/\ell)_\infty((BG_k)_{et})$ and $(\mathbb{Z}/\ell)_\infty \circ \mathrm{Sing.}(BG(\mathbb{C}))$.

9. APPLICATIONS TO TOPOLOGY

In this chapter, we consider two classes of topological applications to the theory we have developed. In Theorem 9.1, we present (a modified version of) D. Sullivan's proof of the Adams Conjecture. This is followed by an infinite loop space version of the Adams Conjecture (Theorem 9.2) whose proof requires the rigidity of the etale topological type and the Bousfield-Kan constructions. Theorem 9.3 presents a method of construction of maps of localized classifying spaces of Lie groups which are not induced by homomorphisms, whereas Theorem 9.5 describes a relativization of this construction to homogeneous spaces. These constructions utilize positive characteristic algebraic geometry.

The reader will observe that all of the applications of this chapter involve the study of simplicial schemes.

In studying the stable homotopy groups of spheres, one employs the *J-homomorphism* $J: O \to \Omega^\infty S^\infty$ given by sending $\alpha: R^n \to R^n$ in O_n to the restriction of α, $J(\alpha): S^{n-1} \to S^{n-1}$ in $\Omega^{n-1}S^{n-1}$. In a series of papers, J.F. Adams determined the order of the image of $J_*: \pi_*(0) \to \pi_*(\Omega^\infty S^\infty) = \pi_*^s(S^0)$ up to 2-torsion [2]. While investigating this J-homomorphism, Adams was led in [1] to a dramatic conjecture (the "Adams Conjecture" verified in Theorem 9.1) concerning the generalized J-homomorphism considered by M.F. Atiyah $J: KO(X) \to J(X)$ (sending a real vector bundle to its associated spherical fibration) for a finite complex X [9], whose solution for spheres completes the determination of $\mathrm{im}(J_*) \subset \pi_*^s(S^0)$.

In an influential paper [61], D. Quillen outlined a proof of the complex K-theory analogue of the Adams Conjecture. This outline employed etale homotopy theory, thus suggesting that the formalism introduced by M. Artin and B. Mazur to study abstract algebraic varieties might be "turned around" so as to utilize algebraic geometry in the study of algebraic topology. D. Quillen's outline was completed by the author in his thesis (cf. [29]); prior to the publication of this complete proof, Quillen in [63] provided an entirely different proof of the Adams Conjecture based on the earlier work of Adams and a technique of "approximating" the orthogonal group by finite groups (see Chapter 12). Independently, D. Sullivan provided a proof of the Adams Conjecture which also used etale homotopy theory ([69]); this proof differed in many respects from Quillen's outline, especially in that it does not involve algebraic varieties in characteristic $p > 0$.

We begin with an outline of Sullivan's proof of the Adams Conjecture. The reader should recall that the Adams operation Ψ^q on (real) K-theory $KO(X)$ is determined by sending a bundle $E \to X$ to the q-th Newton polynomial in the exterior powers $\Lambda^k E \to X$ of $E \to X$ (so that $\psi^q(E \to X)$ is represented by a virtual bundle).

THEOREM 9.1 (Adams Conjecture). *Let* $J : BSO \to BSG$ *represent the (real) J-homomorphism, where* $SO = \cup SO(n)$ *is the infinite special orthogonal group and where* $SG = \cup SG_n$ *with* SG_n *the monoid of oriented self-equivalences of* S^n. *Let* $\Psi^q : BSO \to BSO$ *represent the q-th Adams operation on (real) algebraic K-theory, some* $q > 0$. *Then*

$$J \circ \Psi^q , \quad J : BSO \to BSG$$

determine homotopic maps of $\mathbf{Z}[1/q]$ *localizations*

$$J \circ \Psi^q \sim J : (BSO)_{1/q} \to (BSG)_{1/q} .$$

Proof (sketch). Because BSG is a simply connected space with finite homotopy groups, $(BSG)_{1/q}$ is homotopy equivalent to $\prod_{\ell \nmid q} (BSG)^{\hat{}}_{\ell}$, where $(\ )^{\hat{}}_{\ell} = \mathrm{holim}^{Su}(\ ) \circ (\ )^{\hat{\ell}}$. Because $J \circ \Psi^q$, $J : BSO \to BSG \to (BSG)^{\hat{}}_{\ell}$ factor through $BSO \to (BSO)^{\hat{}}_{\ell}$, it suffices to prove that

$$J \circ \Psi^q, \quad J : (BSO)^\wedge_\ell \to (BSG)^\wedge_\ell$$

are equal in $\mathcal{H}_*$. We recall that there exists a galois automorphism $\sigma \in \mathrm{Gal}(\mathbf{C},\mathbf{Q})$ such that

$$\psi_n = \theta_n \circ \sigma_{et} \circ \theta_n^{-1} : BSO(n)^\wedge_\ell \to ((BSO_{n,\mathbf{C}})_{et})^\wedge_\ell \to ((BSO_{n,\mathbf{C}})_{et})^\wedge_\ell \to (BSO(n))^\wedge_\ell$$

stabilizes (with respect to n) to determine $\Psi^q : (BSO)^\wedge_\ell \to (BSO)^\wedge_\ell$, where θ_n is determined by the two right-most arrows of Proposition 8.8. Consequently, it suffices to prove that

$$J \circ \psi_n, \quad J : (BSO(n))^\wedge_\ell \to (BSG_{n-1})^\wedge_\ell$$

are equal in $\mathcal{H}_*$ for each $n > 0$.

One verifies that maps in $\mathcal{H}_*$ into $(BSG_{n-1})^\wedge_\ell$ are equivalent to fibre homotopy classes of fibrations with fibres $(S^{n-1})^\wedge_\ell$, where $X \to (BSG_{n-1})^\wedge_\ell$ determines the $(S^{n-1})^\wedge_\ell$ fibration given by the pull-back of the universal fibration $B((S^{n-1})^\wedge_\ell, (SG_{n-1})^\wedge_\ell) \to B((SG_{n-1})^\wedge_\ell)$. The fact that the following square

$$\begin{array}{ccc}
BSO(n-1) & \longrightarrow & B(S^{n-1}, SG_{n-1}) \\
\Big\downarrow i & & \Big\downarrow \\
BSO(n) & \xrightarrow{\ J\ } & BSG_{n-1}
\end{array}$$

is homotopy cartesian implies that $i : (BSO(n-1))^\wedge_\ell \to (BSO(n))^\wedge_\ell$ corresponds to $J : (BSO(n))^\wedge_\ell \to (BSG_{n-1})^\wedge_\ell$. Moreover, ψ_{n-1} and ψ_n fit in a homotopy commutative square

$$\begin{array}{ccc}
(BSO(n-1))^\wedge_\ell & \xrightarrow{\ \psi_{n-1}\ } & (BSO(n-1))^\wedge_\ell \\
\Big\downarrow i & & \Big\downarrow i \\
(BSO(n))^\wedge_\ell & \xrightarrow{\ \psi_n\ } & (BSO(n))^\wedge_\ell
\end{array} ;$$

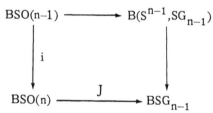

because the horizontal arrows of this square are homotopy equivalences, this square is necessarily homotopy cartesian. We conclude that $i:(BSO(n-1))^{\hat{}}_{\ell} \to (BSO(n))^{\hat{}}_{\ell}$ corresponds to both J and $J \circ \psi_n$ for any $n > 0$, so that $J = J \circ \Psi^q$ in $\mathcal{H}_*$. ∎

We recall that an *Ω-spectrum* $\underset{\sim}{X}$ is a sequence of pointed spaces X_n and maps $i_n : \Sigma X_n \to X_{n+1}$ for $n \geq 0$ such that the adjoint of each i_n is a homotopy equivalence $X_n \to \Omega X_{n+1}$. The J-homomorphism $J : BSO \to BSG$ extends to a map of (connected) Ω-spectra $\underset{\sim}{J} : b\underset{\sim}{SO} \to B\underset{\sim}{SG}$ and the Adams operation $\Psi^q : BSO \to BSO$ determines a map of Ω-spectra $\underset{\sim}{\Psi}^q : (b\underset{\sim}{SO})_{1/q} \to (b\underset{\sim}{SO})_{1/q}$. It is natural to ask whether the maps $\underset{\sim}{J},\ \underset{\sim}{J} \circ \underset{\sim}{\Psi}^q : (b\underset{\sim}{SO})_{1/q} \to (B\underset{\sim}{SG})_{1/q}$ are homotopic as maps of spectra. Unfortunately, this is not the case: $J,\ J \circ \Psi^q : (BSO)_{1/q} \to (BSG)_{1/q}$ are not homotopic as maps of H-spaces, so that $\underset{\sim}{J}$ and $\underset{\sim}{J} \circ \underset{\sim}{\Psi}^q$ cannot have homotopic restrictions to the first delooping of $(BSO)_{1/q}$ ([56]).

Nonetheless, the proof of Theorem 9.1 extends to the following spectrum (or "infinite loop space") version of the complex Adams Conjecture proved by the author in [37]. The proof proceeds by carefully refining Sullivan's homotopy theoretic arguments so that they remain valid in the much more rigid context of spectra. The key step is to interpret a homotopy class of maps of spectra into $(Z/\ell)_\infty(BSG)$ in terms of a geometric structure which arises from algebraic geometry.

THEOREM 9.2. *Let* $\underset{\sim}{J} : b\underset{\sim}{U} \to B\underset{\sim}{SG}$ *denote the* (*complex*) *J-homomorphism of connected Ω-spectra and* $\underset{\sim}{\Psi}^q : (b\underset{\sim}{U})_{1/q} \to (b\underset{\sim}{U})_{1/q}$ *denote the* q*-th Adams operation on the spectrum* $(b\underset{\sim}{U})_{1/q}$ *for some* $q > 0$. *Then*

$$\underset{\sim}{J} \circ \underset{\sim}{\Psi}^q,\ \underset{\sim}{J} : (b\underset{\sim}{U})_{1/q} \to (B\underset{\sim}{SG})_{1/q}$$

are homotopic maps of spectra.

Proof (sketch). An Ω-spectrum $\underset{\sim}{B}$ is determined by a functor $\underline{\underline{B}} : \mathcal{F} \to$ (s. sets$_*$) such that $\prod_{i=1}^{n} p_i : \underline{\underline{B}}(n) \to \underline{\underline{B}}(1)$ is a weak equivalence for each

$n > 1$, where $\mathcal{F}$ is the category of finite pointed sets, $n = \{0,\cdots,n\}$, and $p_i : n \to 1$ satisfies $p_i(i) = 1$, $p_i(j) = 0$ for $j \neq i$. Such "$\mathcal{F}$-spaces" can be obtained by categorical constructions: in particular, both $\underset{\sim}{\mathrm{bU}}$ and $\underset{\sim}{\mathrm{BSG}}$ arise from $\mathcal{F}$-spaces. The proof proceeds by verifying that

$$\underline{\underline{J}},\ \underline{\underline{J}} \circ \underline{\underline{\Psi}}^q : (Z/\ell)_\infty \circ \underline{\underline{\mathrm{bU}}} \to (Z/\ell)_\infty \circ \underline{\underline{\mathrm{BSG}}}$$

are "homotopic" maps of $\mathcal{F}$-spaces for any prime $\ell \not| q$.

Homotopy classes of maps of $\mathcal{F}$-spaces $\underline{\underline{B}} \to (Z/\ell)_\infty \circ \underline{\underline{\mathrm{BSG}}}$ are seen to be in natural one-to-one correspondence with "Z/ℓ-completed S^2-fibrations" over $\underline{\underline{B}}$. The precise definition of such a structure ([37], Definition 7.2) is quite subtle, having been modified repeatedly to permit such a classification theorem for maps into $(Z/\ell)_\infty \circ \underline{\underline{\mathrm{BSG}}}$. The pull-back of the universal Z/ℓ-completed S^2-fibration over $(Z/\ell)_\infty \circ \underline{\underline{\mathrm{BSG}}}$ via the map $\underline{\underline{J}}$ is represented by a structure arising from algebraic geometry (an elaborate version of $\underset{\leftarrow}{\mathrm{holim}} \circ (Z/\ell)_\infty \circ B(A^n_C - \{0\}, (GL_{n,C}))_{et} \to \underset{\leftarrow}{\mathrm{holim}} \circ (Z/\ell)_\infty \circ B(GL_{n,C})_{et}$ which refines the complex analogue of $(BSO(n-1))^{\wedge}_{\ell} \to (BSO(n))^{\wedge}_{\ell}$ in the proof of Theorem 9.1).

The map $\Psi^q : (Z/\ell)_\infty \circ \underline{\underline{\mathrm{bU}}} \to (Z/\ell)_\infty \circ \underline{\underline{\mathrm{bU}}}$ corresponds to a galois action of the algebraic geometry model (an elaborate version of

$$\sigma_{et} : \underset{\leftarrow}{\mathrm{holim}} \circ (Z/\ell)_\infty \circ B(GL_{n,C})_{et} \to \underset{\leftarrow}{\mathrm{holim}} \circ (Z/\ell)_\infty \circ B(GL_{n,C})_{et}\ )$$

and is thus covered by a map of Z/ℓ-completed S^2-fibrations. Consequently, the classification theorem implies that $\underline{\underline{J}} \circ \underline{\underline{\Psi}}^q$ and $\underline{\underline{J}}$ are homotopic maps of $\mathcal{F}$-spaces into $(Z/\ell)_\infty \circ \underline{\underline{\mathrm{BSG}}}$. ∎

D. Quillen once suggested that the constructive aspect of algebraic geometry could provide a valuable approach to various existence problems in algebraic topology. Theorems 9.3 and 9.5 (described below) are examples of the successful application of this philosophy.

We first consider the problem of constructing a map between (localized) classifying spaces of compact, connected Lie groups which is not the classifying map of a homomorphism. In [4], J. F. Adams and Z. Mahmud characterized those maps $BG \dashrightarrow BG$ which could be "defined after finite localization" in terms of "admissible" maps between the universal covering spaces of the associated maximal tori. The localization necessary before Adams and Mahmud were assured of the existence of a map on classifying spaces included the inversion of all primes occurring in the Weyl groups of G and G'. Consequently, the following existence theorem provides a significant sharpening of one aspect of the work of Adams and Mahmud.

One interesting aspect of this theorem is its use of characteristic p algebraic geometry (as did the proof of the complex Adams Conjecture given in [29]). The theorem itself is stated without proof in [36], and the proof given is an easy generalization of that given in [33] for a less general context.

THEOREM 9.3. *Let* $G(\mathbb{C})$ *and* $G'(\mathbb{C})$ *be complex reductive algebraic groups and let* $f: G_F \to G'_F$ *be a homomorphism of associated algebraic groups over* $\operatorname{Spec} F$, *where* F *is the algebraic closure of the prime field* F_p. *Then* f *and a choice of embedding of the Witt vectors of* F *into* $\mathbb{C}$ *(cf. Proposition 8.8) determine a map*

$$\Phi : (BG(\mathbb{C}))_{1/p} \to (BG'(\mathbb{C}))_{1/p}$$

fitting in a homotopy commutative square

$$\begin{array}{ccc} BT(\mathbb{C}) & \xrightarrow{\;B\phi_t\;} & BT'(\mathbb{C}) \\ \downarrow & & \downarrow \\ (BG(\mathbb{C}))_{1/p} & \xrightarrow{\;\Phi\;} & (BG'(\mathbb{C}))_{1/p} \end{array}$$

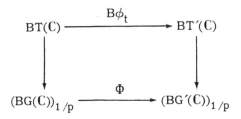

where ϕ_t *is the lifting of the restriction of* $f : G_F \to G'_F$ *to maximal tori of* G_F *and* G'_F.

Proof (Sketch). A choice of isomorphisms $T_F \simeq GL_{1,F}^{\times r}$ and $T'_F \simeq GL_{1,F}^{\times r'}$ determines an isomorphism between the group of $r \times r'$ integer-valued matrices and the group of algebraic group homomorphisms from T_F to T'_F. Thus, we observe that the restriction of f, $f_t : T_F \to T'_F$ fits in a commutative diagram of group schemes

$$\begin{array}{ccccc} T_F & \longrightarrow & T_R & \longleftarrow & T_C \\ {\scriptstyle f_t}\downarrow & & \downarrow & & \downarrow{\scriptstyle \phi_t} \\ T'_F & \longrightarrow & T'_R & \longleftarrow & T'_C \end{array}$$

where R denotes the Witt vectors of F and where $T_C \to T_R$, $T'_C \to T'_R$ are determined by the chosen embedding $R \to \mathbf{C}$.

By Proposition 8.8, f induces a map $\Phi_{\hat{\ell}} : (BG(\mathbf{C}))_{\hat{\ell}} \to (BG'(\mathbf{C}))_{\hat{\ell}}$ restricting to $(B\phi_t)_{\hat{\ell}} : (BT(\mathbf{C}))_{\hat{\ell}} \to (BT'(\mathbf{C}))_{\hat{\ell}}$ for any $\ell \neq p$, where $(\ )_{\hat{\ell}} = \operatorname{holim}^{Su} \circ (\ )^{\ell}$. Because $H^*(BG(\mathbf{C}),\mathbf{Q}) \otimes \mathbf{Q}_\ell = H^*((BG(\mathbf{C}))_{\hat{\ell}},\mathbf{Q})$ and $H^*(BG'(\mathbf{C}),\mathbf{Q}) \otimes \mathbf{Q}_\ell = H^*((BG'(\mathbf{C}))_{\hat{\ell}},\mathbf{Q})$, we conclude that $(B\phi_t)^*$ restricts to

$$\Phi^* : H^*(BG'(\mathbf{C}),\mathbf{Q}) \simeq H^*(BT'(\mathbf{C}),\mathbf{Q})^{W'} \to H^*(BT(\mathbf{C}),\mathbf{Q})^{W} = H^*(BG(\mathbf{C}),\mathbf{Q}).$$

Let $(\ )_{(0)}$ denote localization at (0) (i.e., the rational homotopy type functor). Then $(BG'(\mathbf{C}))_{(0)}$ and $((BG'(\mathbf{C}))_{\hat{\ell}})_{(0)}$ are products of $K(\mathbf{Q},n)$'s and $K(\mathbf{Q}_\ell,n)$'s respectively, so that Φ^* determine a well-defined homotopy class

$$\Phi_{(0)} : (BG(\mathbf{C}))_{(0)} \to (BG'(\mathbf{C}))_{(0)}$$

such that the following squares commute in $\mathcal{H}_*$:

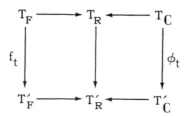

$$\begin{array}{ccccc}
(BG(\mathbb{C}))_{(0)} & \longrightarrow & \prod_{\ell \neq p} ((BG(\mathbb{C}))^{\hat{}}_{\ell})_{(0)} & \longleftarrow & \prod_{\ell \neq p} (BG(\mathbb{C}))^{\hat{}}_{\ell} \\
\Big\downarrow \Phi_{(0)} & & \Big\downarrow \Pi\Phi^{\hat{}}_{\ell} & & \Big\downarrow \Pi\Phi^{\hat{}}_{\ell} \\
(BG'(\mathbb{C}))_{(0)} & \longrightarrow & \prod_{\ell \neq p} ((BG'(\mathbb{C}))^{\hat{}}_{\ell})_{(0)} & \longleftarrow & \prod_{\ell \neq p} (BG'(\mathbb{C}))^{\hat{}}_{\ell}
\end{array}$$

As shown by D. Sullivan (cf. [69], 8.1), this diagram determines $\Phi : (BG(\mathbb{C}))_{1/p} \to (BG'(\mathbb{C}))_{1/p}$ whose homotopy class is well defined on finite skeleta of $(BG(\mathbb{C}))_{1/p}$. To prove the uniqueness (up to homotopy) of this map Φ, we employ the Milnor exact sequence

$$* \to \varprojlim{}^1 \{\mathrm{Hom}_{\mathcal{H}_*}(\Sigma sk_n((BG(\mathbb{C}))_{1/p}),(BG'(\mathbb{C}))_{1/p})\} \to \mathrm{Hom}_{\mathcal{H}_*}((BG(\mathbb{C}))_{1/p},(BG'(\mathbb{C}))_{1/p})$$
$$\to \varprojlim \{\mathrm{Hom}_{\mathcal{H}_*}(sk_n((BG(\mathbb{C}))_{1/p}),(BG'(\mathbb{C}))_{1/p})\} \to * \,.$$

We recall that $H_i((BG(\mathbb{C}))_{1/p})$ and $\pi_i((BG'(\mathbb{C}))_{1/p})$ are finite unless i is even. Therefore, obstruction theory implies that each of the groups $\mathrm{Hom}_{\mathcal{H}_*}(\Sigma sk_n(BG(\mathbb{C}))_{1/p}, (BG'(\mathbb{C}))_{1/p})$ is finite, so that the $\varprojlim{}^1$-term is zero. This implies the uniqueness of Φ. ∎

As a corollary of Theorem 9.3, we obtain the following exceptional equivalences exhibited by the author in [33], the last of which were first exhibited by C. Wilkerson in [71]. These equivalences are those determined by the "*exceptional isogenies*" of algebraic groups.

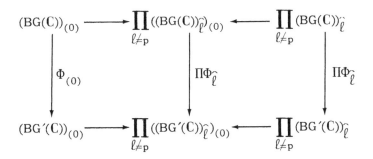

COROLLARY 9.4. a.) *There exists a homotopy equivalence* $(BSO_{2n+1})_{1/2} \to (BSp_n)_{1/2}$ *for any* $n > 0$. b.) *There exists a self equivalence* $\Phi : (BG_2)_{1/3} \to (BG_2)_{1/3}$ *which restricts to* $B\phi_t : BT_2 \to BT_2$, *where* ϕ_t *sends a short root to a long root and a long root to* 3 *times a short root.* c.) *There exists a self equivalence* $\Phi : (BF_4)_{1/2} \to (BF_4)_{1/2}$ *which restricts to* $B\phi_t : BT_4 \to BT_4$, *where* ϕ_t *sends short roots to long*

roots and long roots to 2 *times short roots.* d.) *For any complex reductive Lie group and any prime* $p > 0$, *there exists a self equivalence* $\Phi : (BG(\mathbb{C}))_{1/p} \to (BG(\mathbb{C}))_{1/p}$ *such that* $\Phi^* : H^{2n}(BG(\mathbb{C}),\mathbb{Q}) \to H^{2n}(BG(\mathbb{C}),\mathbb{Q})$ *is multiplication by* p^n *for each* $n \geq 0$. ∎

The comparison theorems of Chapter 8 immediately imply that galois actions determine self equivalences on $\mathbb{Z}/\ell$-completions (but not necessarily on rational cohomology). It would be interesting to understand how the self equivalences of $(\mathbb{Z}/\ell)_\infty \circ \mathrm{Sing.}(BG(\mathbb{C}))$ determined by $\mathrm{Gal}(F, F_p)$ as in Proposition 8.8 depend on the choice of embedding of the Witt vectors into $\mathbb{C}$. Recent work of Z. Wojktowiak [72] appears to answer this question for $G(\mathbb{C}) = GL_n(\mathbb{C})$ and $\ell > n$.

In the following theorem, we present a relativization of Theorem 9.3 concerning *homogeneous spaces* which was proved by the author in [36].

THEOREM 9.5. *Let* $G = G(\mathbb{C})$, $G' = G'(\mathbb{C})$ *be complex reductive Lie groups; let* $G_{\mathbb{Z}}$, $G'_{\mathbb{Z}}$ *be associated Chevalley group schemes over* $\mathrm{Spec}\,\mathbb{Z}$; *and let* $H_{\mathbb{Z}} \subset G_{\mathbb{Z}}$, $H'_{\mathbb{Z}} \subset G'_{\mathbb{Z}}$ *be closed subgroup schemes reductive over* $\mathrm{Spec}\,\mathbb{Z}$. *Let* $f : G_F \to G'_F$ *be a homomorphism of associated algebraic groups over* $\mathrm{Spec}\,F$, *where* F *is the algebraic closure of the prime field* F_p, *such that* f *restricts to* $f_| : H_F \to H'_F$. *Then* f, $f_|$, *and a choice of embedding of the Witt vectors of* F *into* $\mathbb{C}$ *naturally determine a homotopy class of maps*

$$\Phi : (G/H)_{1/p} \to (G'/H')_{1/p} .$$

Proof (Sketch). The map $\Phi_{(0)} : (G/H)_{(0)} \to (G'/H')_{(0)}$ is defined as the unique (up to homotopy–cf. [36]) map fitting in a "map of fibre triples"

$$\begin{array}{ccccc}
(G/H)_{(0)} & \longrightarrow & (BH)_{(0)} & \longrightarrow & (BG)_{(0)} \\
\downarrow & & \downarrow & & \downarrow \\
(G'/H')_{(0)} & \longrightarrow & (BH')_{(0)} & \longrightarrow & (BG')_{(0)}
\end{array}$$

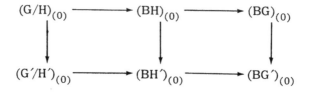

whose middle and right vertical arrows are obtained as in Theorem 9.3. We recall that G/H is naturally homotopy equivalent to $B(G,H,*)$. Moreover, the proof of Proposition 8.8 applies to prove that each map of the chain (determined by a choice of embedding of the Witt vectors R into C)

$$(B(G_F,H_F,*)_{et})^{\wedge}_{\ell} \to (B(G_R,H_R,*)_{et})^{\wedge}_{\ell} \leftarrow (B(G_C,H_C,*)_{et})^{\wedge}_{\ell}$$

$$\leftarrow (B(G_C,H_C,*)_{s.et})^{\wedge}_{\ell} \to (\mathrm{Sing.}(B(G,H,*)))^{\wedge}_{\ell}$$

is induced by a weak equivalence in pro-$\mathcal{H}_*$ and is thus a homotopy equivalence, where $(\ )^{\wedge}_{\ell} = \underleftarrow{\mathrm{holim}}^{\mathrm{Su}} \circ (\ )^{\hat{\ell}}$. The map $\Phi^{\wedge}_{\ell} : (G/H)^{\wedge}_{\ell} \to (G'/H')^{\wedge}_{\ell}$ is that induced by $B(f,f|,*)_{et} : (B(G_F,H_F,*)_{et})^{\wedge}_{\ell} \to (B(G'_F,H'_F,*)_{et})^{\wedge}_{\ell}$ and these equivalences.

The commutativity in $\mathcal{H}_*$ of the following squares

$$\begin{array}{ccccc}
(G/H)_{(0)} & \longrightarrow & \prod_{\ell \neq p} ((G/H)^{\wedge}_{\ell})_{(0)} & \longleftarrow & \prod_{\ell \neq p} (G/H)^{\wedge}_{\ell} \\
\Big\downarrow \Phi_{(0)} & & \Big\downarrow \Pi\Phi^{\wedge}_{\ell} & & \Big\downarrow \Pi\Phi^{\wedge}_{\ell} \\
(G'/H')_{(0)} & \longrightarrow & \prod_{\ell \neq p} ((G'/H')^{\wedge}_{\ell})_{(0)} & \longleftarrow & \prod_{\ell \neq p} (G'/H')^{\wedge}_{\ell}
\end{array}$$

follows from the uniqueness of $\Phi^{\wedge}_{\ell} : ((G/H)^{\wedge}_{\ell})_{(0)} \to ((G'/H')^{\wedge}_{\ell})_{(0)}$ fitting in a "map of fibre triples." Consequently, the finite dimensionality of G/H implies that these squares determine a unique homotopy class of maps $\Phi : (G/H_{1/p} \to (G'/H')_{1/p}$. ■

As in Corollary 9.4, one obtains explicit examples of exceptional maps using exceptional isogenies. We refer the interested reader to [36].

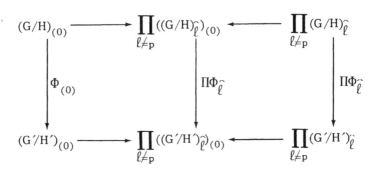

10. COMPARISON OF GEOMETRIC AND HOMOTOPY THEORETIC FIBRES

In this chapter, we relate the cohomology of the homotopy theoretic fibre of the etale topological type of a map $f: X. \to Y.$ of simplicial schemes, $\mathrm{fib}(f_{et})$, to the cohomology of the etale topological type of the scheme theoretic fibre, $(X. \times_{Y.} y)_{et}$. In the special case of a map of schemes and in the context of etale homotopy types, this comparison constituted the central part of the author's thesis [29]. One satisfying (and new) aspect of the presentation of this chapter is that it gives a more natural description of the map inducing the comparison even in this special context. The applications of etale homotopy theory discussed in Chapter 11 require the comparison for maps of simplicial schemes (first obtained for etale homotopy types in [34]); the author's original application to the proof of the Adams Conjecture requires only a comparison for maps of schemes.

Definition 10.1 introduces various homotopy theoretic fibres associated to $f: X. \to Y.$. We must consider several definitions because the crucial comparison involves special maps of (non-rigid) hypercoverings; thus, the necessity of Propositions 10.2 and 10.3 relating $\mathrm{fib}(f_{et})$ to the homotopy theoretic fibre associated to such special maps. Proposition 10.4 provides the critical insight: the cohomology of the "naive fibre" associated to special maps is naturally isomorphic to the derived functors of $f_*: \mathrm{AbSh}(X) \to \mathrm{AbSh}(Y)$. This permits the identification of the Leray spectral sequence for f_* with a spectral sequence arising from homotopy theory. Theorem 10.7 presents the comparison theorem, whose proof is based on a comparison of Leray and Serre spectral sequences. Examples are given in Corollaries 10.8 and 10.9.

DEFINITION 10.1. Let $f: X.\,, x \to Y.\,, y$ be a pointed map of locally noetherian simplicial schemes. Define $HR(f)$ to be the category whose objects are maps $g: U.. \to V..$ of pointed hypercoverings over f and whose maps are equivalence classes of commutative squares $g \to g'$ of pointed maps over f, where two maps $g \rightrightarrows g'$ are equivalent if they are related by a pointed simplicial homotopy $g \otimes \Delta[1] \to g'$ over f (cf. Definition 3.3). Define $HRR(f)$ to be the (left directed) subcategory of $HR(f)$ whose objects $g: U.. \to V..$ satisfy the condition that g be a map of rigid hypercoverings over f and whose maps are commutative squares of rigid hypercoverings over f. Define $Sp(f)$ to be the full subcategory of $HR(f)$ whose objects $g: U.. \to V..$ satisfy the condition that $U.. \to V.. \times_{Y.} X.$ be special (cf. proof of Proposition 3.4).

Define the *homotopy fibres* $fib(f_{ht})$, $fib(f_{et})$, and $fib(f_{sp})$ by

$$fib(f_{ht}) = \{(\pi(\Delta g)^{\sim})^{-1}(y);\ g \in HR(f)\}$$

$$fib(f_{et}) = \{(\pi(\Delta g)^{\sim})^{-1}(y);\ g \in HRR(f)\}$$

$$fib(f_{sp}) = \{(\pi(\Delta g)^{\sim})^{-1}(y);\ g \in Sp(f)\}$$

where $\pi(\Delta g)^{\sim}: \pi(\Delta U..)^{\sim} \to \pi(\Delta V..)$ is the mapping (Kan) fibration of $\pi(\Delta g)$ as constructed in ([42], VI. 5.5.1) and $y \in \pi(V_{0,0})$ is the base point. ■

The naturality of the (homotopy) left equalizer of a pair of maps between hypercoverings as constructed in the proof of Proposition 3.4 shows that $HR(f)$ has left equalizers as well as products; thus, $HR(f)$ is left filtering. Similarly, the left equalizer in $HR(f)$ of a pair of maps $g \rightrightarrows g'$ in $Sp(f)$ is again in $Sp(f)$, as can be readily checked using Lemma 1.3 of [29]; and the product in $HR(f)$ of objects in $Sp(f)$ is again in $Sp(f)$, so that $Sp(f)$ is also left filtering. Thus, $fib(f_{ht})$ and $fib(f_{sp})$ are objects of pro-$\mathcal{H}_*$ and $fib(f_{et})$ is an object of pro-(s. sets$_*$).

The next proposition introduces the natural map between geometric and homotopy theoretic fibres for both etale topological and homotopy types. If $f: X. \to Y.$ is a map of simplicial schemes and $y: \mathrm{Spec}\,\Omega \to Y_0$

is a geometric point, we define $(X. \times_{Y.} y)_n$ to be $X_n \times_{Y_n} \operatorname{Spec} \overline{k(y)}$, where $\overline{k(y)}$ is the separable closure inside Ω of the residue field of the image Zariski point of $\sigma_0^n y : \operatorname{Spec} \Omega \to Y_n$ (so that the residue fields of $X. \times_{Y.} y$ are subfields of Ω).

PROPOSITION 10.2. *Let* $f : X. \to Y.$ *be a pointed map of locally noetherian simplicial schemes, and let* $i : X. \times_{Y.} y \to X.$ *be the geometric fibre of* f. *The functors*

$$i^* \circ s : HR(f) \to HR(X.) \to HR(X. \times_{Y.} y) ,$$

$$i^* \circ s : HRR(f) \to HRR(X.) \to HRR(X. \times_{Y.} y)$$

together with the natural closed immersions $i^*(U..) \to U.. = s(g : U.. \to V..)$ *determine natural maps from the geometric to the homotopy theoretic fibres*

$$(X. \times_{Y.} y)_{ht} \to \operatorname{fib}(f_{ht}), \quad (X. \times_{Y} y)_{et} \to \operatorname{fib}(f_{et})$$

which fit in a natural commutative square in pro-$\mathcal{H}_*$

$$\begin{array}{ccc} (X. \times_{Y.} y)_{ht} & \longrightarrow & \operatorname{fib}(f_{ht}) \\ \downarrow & & \downarrow \\ (X. \times_{Y.} y)_{et} & \longrightarrow & \operatorname{fib}(f_{et}) \end{array}$$

such that $(X. \times_{Y.} y)_{ht} \to (X. \times_{Y.} y)_{et}$ *is weak equivalence in* pro-$\mathcal{H}_*$. *Moreover,* $\operatorname{fib}(f_{ht}) \to \operatorname{fib}(f_{et})$ *is also a weak equivalence in* pro-$\mathcal{H}_*$ *if either* $\pi_0((X. \times_{Y.} y)_{ht}) \to \pi_0(\operatorname{fib}(f_{ht}))$ *is a bijection of finite sets or if* f *is a map of schemes.*

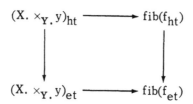

Proof. Let $W.$ denote $X. \times_{Y.} y$. For each connected component $W.^{\alpha}$ of $W.$, choose a geometric point x_α of W_0^α (with $x_\alpha = x$ for the distinguished component). As in the proof of Corollary 6.3, the forgetful function $HR(W_\cdot, \{x_\alpha\}) \to HR(W_\cdot, x)$ is left final. Consequently, $((W_\cdot, x)_{ht}) \to (W_\cdot)_{et}$ is isomorphic in pro-$\mathcal{H}_*$ to $(W_\cdot, \{x_\alpha\})_{ht} \to (W_\cdot)_{et}$, which is a disjoint union of maps of pro-$\mathcal{H}_{*c}$ each of which is a weak equivalence by Corollary 6.3. Thus, $((W_\cdot, x)_{ht}) \to (W_\cdot)_{et}$ is a weak equivalence in pro-$\mathcal{H}_*$.

In a similar fashion, $HR(f, \{x_\alpha\}) \to HR(f)$ is left final, where $HR(f, \{x_\alpha\})$ consists of pointed maps $g: U_{\cdot\cdot} \to V_{\cdot\cdot}$ with the added structure of geometric points u_α above each x_α, provided that $\{x_\alpha\}$ is finite. If

$$\pi_0((X_\cdot \times_{Y_\cdot} y)_{ht}) \to \pi_0(\mathrm{fib}(f_{ht}))$$

is a bijection of finite sets, then $\mathrm{fib}(f_{ht})$ is isomorphic to a disjoint union of objects in pro-$\mathcal{H}_{*c}$ (using $HR(f, \{x_\alpha\})$ as an indexing category). Using the homotopy sequences for

$$\mathrm{fib}(f_{ht}) \to (X_\cdot)_{ht} \to (Y_\cdot)_{ht} \quad \text{and} \quad \mathrm{fib}(f_{et}) \to (X_\cdot)_{et} \to (Y_\cdot)_{et}$$

we conclude that $\mathrm{fib}(f_{ht}) \to \mathrm{fib}(f_{et})$ induces a bijection on connected components and is isomorphic to a disjoint union of maps in pro-$\mathcal{H}_{*c}$. Applying the map between homotopy sequences for each component of $\mathrm{fib}(f_{ht}) \to \mathrm{fib}(f_{et})$, we conclude using Theorem 6.2 that this map is a weak equivalence in pro-$\mathcal{H}_*$.

Finally, if f is a map of schemes, the proof of Proposition 4.5 applies to prove that f_{ht} is isomorphic to f_{et} in pro-$\mathcal{H}_*^2$, so that $\mathrm{fib}(f_{ht})$ is isomorphic to $\mathrm{fib}(f_{et})$ in pro-$\mathcal{H}_*$. ∎

In the following proposition, we prove that the natural map $\mathrm{fib}(f_{ht}) \to \mathrm{fib}(f_{sp})$ is an isomorphism in pro-$\mathcal{H}_*$. This is somewhat surprising at first glance, because the inclusion $Sp(f) \to HR(f)$ inducing this map is not left final.

PROPOSITION 10.3. *Let* $f: X_\cdot \to Y_\cdot$ *be a pointed map of locally noetherian simplicial schemes. The inclusion* $j: Sp(f) \to HR(f)$ *induces an isomorphism*

$$\mathrm{fib}(f_{ht}) \xrightarrow{\sim} \mathrm{fib}(f_{sp}) \quad \text{in pro-}\mathcal{H}_* .$$

Moreover, the composition of this isomorphism with the natural map $(X. \times_{Y.} y)_{ht} \to \mathrm{fib}(f_{ht})$ *of Proposition 10.2,*

$$(X. \times_{Y.} y)_{ht} \to \mathrm{fib}(f_{sp}) ,$$

is determined by the functor

$$\eta : \mathrm{Sp}(f) \to \mathrm{HR}(X. \times_{Y.} y, y)$$

sending $g : U.. \to V..$ *to* $U.. \times_{V..} v$ *where* $v : \mathrm{Spec}\,\Omega \to V_{0,0}$ *is the chosen geometric point of* $V..$ *over* y.

Proof. We define a functor $\psi : \mathrm{HR}(f) \to \mathrm{Sp}(f)$ by sending $g : U.. \to V..$ to $\psi(g) : E(g) \to V..$, where $E(g)$ is the (homotopy) left equalizer (as constructed in Proposition 3.4) of the two maps

$$pr_1, g \circ pr_2 : V.. \times_{Y.} U.. \to V..$$

and where $\psi(g)$ is the composition

$$pr_1 \circ pr : E(g) \to V.. \times_{Y.} U.. \to V..$$

($E(g) \to V.. \times_{Y.} X$ induced by $\psi(g)$ is the composition of the special maps $E(g) \to V.. \times_{Y.} U..$ and $pr_1 : V.. \times_{Y.} U.. \to V.. \times_{Y.} X.$). We define $\psi' : \mathrm{HR}(f) \to \mathrm{HR}(f)$ by sending $g : U.. \to V..$ to

$$\psi'(g) = g \circ pr_2 \circ pr : E(g) \to V.. \times_{Y.} U.. \to U.. \to V.. .$$

Then there are natural transformations $1 \to j \circ \psi$, $1 \to \psi'$, and $\psi' \to 1$. Moreover, $\psi(g)$ and $\psi'(g)$ are related by a canonical homotopy

$$\theta(g) : E(g) \times \Delta[1] \to V.. ,$$

so that there are canonical isomorphisms

$$\pi(\Delta\psi(g))^{\sim} \xrightarrow{\simeq} \pi(\Delta\psi'(g))^{\sim} \quad \text{in } \mathcal{H}^2_*$$

and

$$(\pi(\Delta\psi(g))^{\sim})^{-1}(y) \xrightarrow{\simeq} (\pi(\Delta\psi'(g))^{\sim})^{-1}(y) \quad \text{in } \mathcal{H}_* .$$

We define $\text{fib}(f_{sp}) \to \text{fib}(f_{ht})$ via the functor ψ and the natural maps

$$(\pi(\Delta\psi(g))^{\sim})^{-1}(y) \xrightarrow{\simeq} (\pi(\Delta\psi'(g))^{\sim})^{-1}(y) \longrightarrow (\pi(\Delta g)^{\sim})^{-1}(y) .$$

The reader can easily verify that $\text{fib}(f_{sp}) \to \text{fib}(f_{ht})$ is homotopy inverse to $\text{fib}(f_{ht}) \to \text{fib}(f_{sp})$ using the fact that $g \to \psi(g) \to \theta(g)$ and $g \to \psi'(g) \to \theta(g)$ are homotopic.

To verify the second assertion, we observe that the composition

$$(X. \times_{Y.} y)_{ht} \to \text{fib}(f_{ht}) \to \text{fib}(f_{sp})$$

is given by the functor

$$i^* \circ s : \text{Sp}(f) \to \text{HR}(X. \times_{Y.} y, y)$$

sending $g : U.. \to V..$ to $U.. \times_{Y.} y$. The proof of the assertion follows from the existence of natural inclusions $U.. \times_{V..} v \to U.. \times_{Y.} y$ in $\text{HR}(X. \times_{Y.} y, y)$ determining a natural transformation $\eta \to i^* \circ s$. ∎

The next proposition explains our introduction of Sp(f) and our consideration of $\text{fib}(f_{sp})$. The proposition originally appeared as Proposition 2.4 of [29].

PROPOSITION 10.4. *Let* $f : X \to Y$ *be a pointed map of locally noetherian schemes and let* $f_{sp}^{-1}(y) \in \text{pro-}\mathcal{H}_*$ *be defined by*

$$f_{sp}^{-1}(y) = \{\pi(U. \times_{V.} v_0);\ g : U. \to V. \ \textit{in}\ \text{Sp}(f)\} = \{\pi(g)^{-1}(y)\,;\ g \in \text{Sp}(f)\}$$

(*where* $v_0 \to V.$ *is the inclusion of the distinguished component of* V_0). *For any locally constant abelian sheaf* M *on* Et(X), *there exists a natural isomorphism*

$$H^*(f_{sp}^{-1}(y),M) \xrightarrow{\sim} (R^*f_*M)_y$$

which fits in a commutative square

$$\begin{array}{ccc} H^*(f_{sp}^{-1}(y),M) & \xrightarrow{\sim} & (R^*f_*M)_y \\ \downarrow & & \downarrow \\ H^*((X_y)_{ht},i^*M) & \xrightarrow{\sim} & H^*(X_y,i^*M) \end{array}$$

where $i: X_y = X \times_Y y \to X$ *is the geometric fibre of* f, *the left arrow is induced by the map* $(X_y)_{ht} \to f_{sp}^{-1}(y)$ *given by* η *of Proposition 10.3, the right arrow is the canonical base change map, and the bottom arrow is the isomorphism of Proposition 5.9.*

Proof. The functor $\eta : Sp(f) \to HR(X_y,y)$ of Proposition 10.3 induces the following map of spectral sequences (whose colimits are indexed by $g: U. \to V.$ in Sp(f) and W. in $HR(X_y,y)$) because $U_p \times_{V_p} v_0 \to X \times_Y v_0$ is a hypercovering

$$\begin{array}{ccc} E_1^{p,q} = \operatorname{colim} H^q(U_p \times_{V_p} v_0,M) & \Longrightarrow & \operatorname{colim} H^{p+q}(U. \times_{V.} v_0,M) \\ \downarrow & & \\ 'E_1^{p,q} = \operatorname{colim} H^q(W_p,i^*M) & \Longrightarrow & \operatorname{colim} H^{p+q}(W.,i^*M) . \end{array}$$

By Proposition 3.7,

$$\operatorname{colim} H^*(U. \times_{V.} v_0,M) = \operatorname{colim} H^*(X \times_Y v_0,M) = (R^*f_*M)_y$$

and

$$\operatorname{colim} H^*(W.,i^*M) = H^*(X_y,i^*M) .$$

The map on abutments is induced by the closed immersions $X_y \to X \times_Y v_0$ and is therefore the base change map. As seen in Theorem 3.8, the $'E^{p,q}$ spectral sequence collapses at the E_2-level with

$$E_2^{*,0} \xrightarrow{\sim} H^*((X_y)_{ht}, i^*M)$$

as in Proposition 5.9. Similarly, the $E^{p,q}$ spectral sequences collapses at the E_2-level (use the proof of Theorem 3.8 with the presheaf G_3 replaced by $H^q(\ ,M)$) and $E_2^{*,0} = \operatorname{colim} H^*(\pi(U. \times_{V.} v),M))$. Let $H^*(f_{sp}^{-1}(y),M) \xrightarrow{\sim} (R^*f_*M)_y$ be the edge isomorphism $E_2^{*,0} \xrightarrow{\sim} E_\infty^*$. Then the asserted commutative diagram is a consequence of the naturality of the edge homomorphism. ∎

Although Proposition 10.4 has been presented only for maps of schemes, its applicability to maps of simplicial schemes is a consequence of the following lemma.

LEMMA 10.5. *Let* $f: X. \to Y.$ *be a pointed map of locally noetherian simplicial schemes with the property that each* X_n *and each* Y_n *is connected, and let* M *be a locally constant abelian sheaf on* Et(X.). *If the map* $(X. \times_{Y.} y)_{ht} \to \mathrm{fib}(f_{sp})$ *of Proposition 10.3 induces isomorphisms for each* $n \geq 0$

$$H^*(\mathrm{fib}((f_n)_{sp}),M_n) \xrightarrow{\sim} H^*((X_n \times_{Y_n} y)_{ht}, i^*M_n),$$

then this map also induces an isomorphism

$$H^*(\mathrm{fib}(f_{sp}),M) \xrightarrow{\sim} H^*((X. \times_{Y.} y)_{ht}, i^*M).$$

Proof. For any $g: U.. \to V..$ in HR(f), let $\pi(g)^{\sim}: \pi(U..)^{\sim} \to \pi(V..)$ be defined by the condition that $(\pi(g)^{\sim})_{n.}$ be the mapping fibration of $\pi(g_n)$ for each $n \geq 0$. Then $\pi(\Delta g) \to \pi(\Delta g)^{\sim}$ factors through a natural map $\Delta(\pi(g)^{\sim}) \to \pi(\Delta g)^{\sim}$: this is proved by observing for any map of bi-simplicial sets $h: S.. \to T..$ and any $\varepsilon_i: \Lambda[n] \to \Delta[n]$ (the inclusion of $sk_{n-1}\Delta[n]$ minus the i-th face into $\Delta[n]$) that a map $\varepsilon_i \to h_{n.}$ naturally determines a map $\varepsilon_i \to \Delta(h)$. Moreover, because $\pi(U_{n.})$ and $\pi(V_{n.})$ are connected for each $n \geq 0$, Theorem B.4 of [12] implies that

$$\Delta(\pi(g)^{\sim})^{-1}(y) \to (\pi(\Delta g)^{\sim})^{-1}(y)$$

is a weak equivalence.

Consequently, the natural map $f_{sp}^{-1}(y) \to \mathrm{fib}(f_{sp})$ factors through the natural weak equivalence

$$\mathrm{fib}'(f_{sp}) = \{\Delta(\pi(g)^{\sim})^{-1}(y);\ g \in Sp(f)\} \to \mathrm{fib}(f_{sp}) .$$

The functor $\eta: Sp(f) \to HR(X. \times_{Y.} y,y)$ of Proposition 10.2 induces a map of spectral sequences

$$\begin{array}{ccc} E_1^{p,q} = H^q((X_p \times_{Y_p} y)_{ht},M_p) & \Rightarrow & H^{p+q}((X. \times_{Y.} y)_{ht},M) \\ & & \downarrow \\ 'E_1^{p,q} = H^q(\mathrm{fib}((f_p)_{sp}),M_p) & \Rightarrow & H^{p+q}(\mathrm{fib}'(f_{sp}),M) \end{array}$$

in view of the fact that $HR(X. \times_{Y.} y,y) \to HR(X_p \times_{Y_p} y,y)$ and $Sp(f) \to Sp(f_p)$ are left final (cf. Proposition 3.4). The hypothesis of the lemma now implies that this map is an isomorphism of spectral sequences. ∎

Thanks to the preceding discussion, the study of

$$H^*(\mathrm{fib}(f_{et}),M) \to H^*((X. \times_{Y.} y)_{et}, i^*M)$$

is reduced to the study of a map of schemes $f: X \to Y$ and the map in cohomology induced by $(X_y)_{ht} \to \mathrm{fib}(f_{sp})$ (if $H^0(\mathrm{fib}(f_{sp}),M) \xrightarrow{\simeq} H^0(X. \times_{Y.} y, i^*M)$, then $\pi_0((X. \times_{Y.} y)_{ht}) \to \pi_0(\mathrm{fib}(f_{ht}))$ is a bijection). Moreover, this map factors as

$$(X_y)_{ht} \to f_{sp}^{-1}(y) \to \mathrm{fib}(f_{sp}) ,$$

with the map in cohomology induced by $(X_y)_{ht} \to f_{sp}^{-1}(y)$ well understood by Proposition 10.4.

The next step in our comparison is given by Theorem 3.9 of [29]. This theorem asserts that the Leray spectral sequence for $f: X \to Y$ can be identified with the Serre spectral sequence for

$$\{\pi(g): \pi(U.) \to \pi(V.);\ g \in Sp(f)\}$$

(this should be compared with Proposition 10.4).

The proof of our comparison theorem (Theorem 10.7) consists of studying the map from the Leray spectral sequence for $f: X \to Y$ to the Serre spectral sequence for

$$\{\pi(g)^{\sim} : \pi(U.)^{\sim} \to \pi(V.);\ g \in Sp(f)\}$$

involving $f_{sp}^{-1}(y) \to fib(f_{sp})$. In order to compare spectral sequences, we require the following lemma which enables us to trivialize the action $\pi_1((Y)_{ht})$ on $H^*(f_{sp}^{-1}(y),M)$ and $H^*(fib(f_{sp}),M)$.

LEMMA 10.6. *Let* $g: Z' \to Z$ *be a pointed principal* G-*fibration of locally noetherian schemes for some finite group* G. *If* $V. \to Z$ *is a pointed hypercovering with the property that* $V_0 \to Z$ *factors through* g, *then* $\pi(V. \times_Z Z') \to \pi(V.)$ *is a pointed principal* G-*fibration of simplicial sets. Moreover, the map in* pro-$\mathcal{H}_*$

(10.6.1) $(Z'.)_{ht} \to \{\pi(V. \times_{Z.} Z'.);\ V. \in HR(Z,z),\ V_0 \to Z$ *factors through* $g\}$

is isomorphic to a disjoint union of weak equivalences in pro-$\mathcal{H}_{*c}$.

Proof. Because $Z' \times_Z Z' \simeq \coprod_{g \in G} Z'$, $V_n \times_Z Z' \to V_n$ is isomorphic to $\coprod_{g \in G} V_n \to V_n$ whenever $V_0 \to Z$ factors through $Z' \to Z$. This implies that $\pi(V. \times_Z Z') \to \pi(V.)$ is a principal G-fibration of simplicial sets whenever $V_0 \to Z$ factors through g. As argued in the proof of Proposition 10.2, we may assume Z' is connected, so that the map (10.6.1) is a map in pro-$\mathcal{H}_{*c}$. Corollary 5.7 implies that (10.6.1) induces an isomorphism of fundamental pro-groups. To prove that (10.6.1) induces an isomorphism in cohomology with any local coefficient system M as coefficients, we observe that the colimit (indexed by $V. \in HR(Z,z)$) spectral sequence

$$E_1^{p,q} = \operatorname{colim} H^q(V_p \times_Z Z',M) \Rightarrow H^{p+q}(Z',M)$$

degenerates at the E_2-level: $E_1^{p,q} = 0$ for $q > 0$ since

$$\operatorname*{colim}_{y \to v} H^q(v \times_Z Z',M) = 0 \quad \text{for} \quad q > 0$$

and any geometric point y of Z (cf. proof of Theorem 3.8 with G_3 replaced by $H^q(\ \times_Z Z',M)$). ■

We now present our basic comparison theorem, a generalization to simplicial schemes of [29], Theorem 4.5.

THEOREM 10.7. *Let* $f: X. \to Y.$ *be a pointed map of simplicial schemes such that each* X_n *is connected, each* Y_n *is connected and noetherian, each* $\pi_1((Y_n)_{ht})$ *is profinite, and* $\pi_0(X. \times_{Y.} y)$ *is finite. Let* M *be a locally constant abelian sheaf on* $Et(X.)$ *such that for each* $n,q \geq 0$, $R^q f_{n*} M_n$ *is locally constant on* $Et(Y_n)$ *and the base change map*

$$(R^q f_{n*} M_n)_y \to H^q(X_n \times_{Y_n} y, i^* M_n)$$

is an isomorphism. Then the natural map $(X. \times_{Y.} y)_{et} \to fib(f_{et})$ *of Proposition 10.2 induces an isomorphism in cohomology*

$$H^*(fib(f_{et}),M) \simeq H^*((X. \times_{Y.} y)_{et}, i^*M) .$$

Proof. The hypothesis $(R^q f_{n*} M_n)_y \xrightarrow{\sim} H^q(X_n \times_{Y_n} y, i^* M_n)$ implies that $X_n \times_{Y_n} y \to (f_n)_{sp}^{-1}(y)$ induces an isomorphism

$$H^*((f_n)_{sp}^{-1}(y), M_n) \xrightarrow{\sim} H^*((X_n \times_{Y_n} y)_{ht}, i^* M_n)$$

by Proposition 10.4. Consequently, as discussed prior to Lemma 10.6, it suffices to prove that $(f_n)_{sp}^{-1}(y) \to fib((f_n)_{sp})$ induces an isomorphism

$$H^*(\mathrm{fib}((f_n)_{sp}),M) \to H^*((f_n)_{sp}^{-1}(y),M) \quad \text{for each} \quad n \geq 0 .$$

We let C denote the left directed category of connected, pointed, principal G-fibrations $Y_n' \to Y_n$, and we let $f_n' : X_n' \to X_n$ denote the pull-back of f_n by $Y_n' \to Y_n$. The maps $(f_n')_{sp} \to (f_n')_{sp}^{-}$ indexed by $Y_n' \to Y_n$ in C determine maps of spectral sequences inducing isomorphisms on abutments

$$\begin{array}{ccc} E_2^{p,q}(Y_n' \to Y_n) = & H^p((Y_n')_{ht}, H^q(\mathrm{fib}(f_n'),M)) & \Rightarrow H^{p+q}(X_n',M) \\ & \downarrow & \\ {}'E_2^{p,q}(Y_n' \to Y_n) = & H^p((Y_n')_{ht}, H^q((f_n')_{sp}^{-1}(y),M)) & \Rightarrow H^{p+q}(X_n',M) . \end{array}$$

The hypothesis that $R^q f_{n*} M_n$ is locally constant on $Et(Y_n)$ implies that $R^q f'_{n*} M_n$ (the restriction of $R^q f_{n*} M_n$) is locally constant on $Et(Y_n')$; the fact that ${}'E^{p,q}$ can be identified with the Leray spectral sequence implies that the coefficients for the cohomology groups of $(Y_n')_{ht}$ are a local coefficient system in the case of ${}'E_2^{p,q}$ as well as in the case of $E_2^{p,q}$. (We have implicitly used the noetherian hypothesis on Y_n to obtain the E_2-terms in the above form by taking a direct limit of coefficients; for more details, see [29], 4.3.)

The map $f_n' \to f_n$ clearly induces an isomorphism on geometric fibres

$$X_n' \times_{Y_n'} y' \xrightarrow{\sim} X_n \times_{Y_n} y .$$

Moreover, Lemma 10.6 implies that $\mathrm{fib}(f_n') \to \mathrm{fib}(f_n)$ is an isomorphism in pro-$\mathcal{H}_*$. We take the colimit (indexed by C^{op}) of the maps

$$E^{p,q}(Y_n' \to Y_n) \to {}'E^{p,q}(Y_n' \to Y_n)$$

so that colim ${}'E_2^{p,q}$ is cohomology with trivial coefficients (the action of $\pi_1(Y_n)$ on $R^q f_{n*} M_n$ is trivialized by some $Y_n' \to Y_n$); moreover, colim $E_2^{p,q}$ is also cohomology with trivial coefficients because $\pi_1((Y_n)_{ht})$ is pro-finite and the action of $\pi_1((Y_n)_{ht})$ on $H^*(\mathrm{fib}(f_n),M)$ is continuous.

Comparing these colimit spectral sequences, we conclude that the map

$$H^*(\mathrm{fib}((f_n)_{sp}),M) \xrightarrow{\simeq} \operatorname{colim} H^*(\mathrm{fib}((f_n')_{sp}),M)$$

$$\longrightarrow \operatorname{colim} H^*((f_n')_{sp}^{-1}(y),M) \xrightarrow{\simeq} H^*((f_n)_{sp}^{-1}(y),M)$$

is an isomorphism as required. ∎

The proper smooth base change theorem ([7], XVI.2.2) asserts that $f: X \to Y$ and $M \in \mathrm{AbSh}(X)$ as below satisfy the hypotheses of Theorem 10.7.

COROLLARY 10.8. *Let* $f: X \to Y$ *be a proper, smooth, pointed map of connected, noetherian schemes and let* M *be a locally constant, constructible sheaf on* $\mathrm{Et}(X)$ *with stalks of order relatively prime to the residue characteristics of* Y. *If* $\pi_1(Y_{et})$ *is pro-finite, then the natural map* $(X_y)_{et} \to \mathrm{fib}(f_{et})$ *induces an isomorphism*

$$H^*(\mathrm{fib}(f_{et}),M) \xrightarrow{\simeq} H^*((X_y)_{et},M) . \blacksquare$$

In the proof of the Adams Conjecture presented in [29], the following special case of Theorem 10.7 was required (cf. [29], 5.3). An independent proof that $\mathrm{fib}(f_{ht})$ is weakly ℓ-equivalent to a sphere has been provided by D. Cox in [21] (using the existence of a "Thom isomorphism"). The fact that the map $f: X \to Y$ and the abelian group of Corollary 10.9 satisfy the hypotheses of Theorem 10.7 is verified in Proposition 5.2 of [29].

COROLLARY 10.9. *Let* Y *be a pointed, connected, geometrically unibranched noetherian scheme and let* E *be a coherent, locally free sheaf on* Y. *Let* $V(E) = \mathrm{Sym}(E^{\vee})$ *be defined locally as the spectrum of the symmetric algebra of the dual of* E *over* O_Y *and let* $f: X \to Y$ *denote the structure map* $V(E) - o(Y) \to Y$, *where* $o: Y \to V(E)$ *is the* 0*-section. For any finite abelian group* A *of order relatively prime to the residue characteristics of* Y, *the natural map* $(X_y)_{et} \to \mathrm{fib}(f_{et})$ *induces an isomorphism in cohomology*

$$H^*(\mathrm{fib}(f_{et}),A) \xrightarrow{\simeq} H^*((X_y)_{et},A) . \blacksquare$$

11. APPLICATIONS TO GEOMETRY

In this chapter, we present four applications of etale homotopy theory to geometry. Theorem 11.1 is a result of P. Deligne and D. Sullivan asserting that a "flat" vector bundle can be trivialized by passing to a finite covering space of the base. The next application, Corollary 11.3, is a necessary and sufficient condition due to M. Artin and J.-L. Verdier for a real algebraic variety to have a real point; the proof we present is based on a result of D. Cox concerning the etale homotopy type of a real variety. Theorem 11.5 presents a long exact homotopy sequence associated to certain maps ("geometric fibrations") of schemes and their geometric fibres. Finally, we verify in Theorem 11.7 that a smooth algebraic variety over an algebraically closed field k has a base of etale neighborhoods whose etale homotopy types have ℓ-completions which are $K(\pi,1)$'s for ℓ a prime invertible in k.

We recall that a complex vector bundle over a pointed, connected base has a discrete structure group if and only if it arises from a complex representation of the fundamental group of the base (as the bundle associated to the corresponding local coefficient system whose fibres are complex vector spaces). If the base of a finite dimensional, complex vector bundle is a manifold, then the bundle has discrete structure group if and only if the bundle admits an integrable connection (in which case, the bundle is said to be "*flat*"). The conclusion of the following theorem describing how such a vector bundle can be trivialized (but not addressing the question of trivializing the associated representation of the fundamental group of the base) is a homotopy theoretic statement amenable to etale homotopy theoretic techniques.

THEOREM 11.1 (Deligne-Sullivan [25]). *Let* T *be a finite simplicial complex and let* $p: E \to T$ *be a finite dimensional, complex vector bundle with discrete structure group. Then there is a finite, surjective covering space* $T' \to T$ *such that the pull-back* $p': E' \to T'$ *of* p *to* T' *is a trivial vector bundle.*

Proof (Sketch). We are easily reduced to the case T is pointed and connected. Let $\rho: \pi_1(T,t) \to GL_n(\mathbb{C})$ be a representation determining $p: E \to T$. Because $\pi_1(T,t)$ is finitely generated, we may find a subring A of $\mathbb{C}$ of finite type over $\mathbb{Z}$ such that ρ is given by $\rho: \pi_1(T,t) \to GL_n(A)$. Deligne and Sullivan show that the classification map for p, $f: T \to BU_n$, can be represented up to homotopy by a map $g_A(\mathbb{C}): X_A(\mathbb{C})^{top} \to Grass_{N+n,n;A}(\mathbb{C})^{top}$, where

$$g_A : X_A \to Grass_{N+n,n;A}$$

is a map of schemes defined over A.

Let m and m' be maximal ideals of A whose (finite) residue fields $A/m = k$ and $A/m' = k'$ have distinct characteristics. An argument similar to that of the proof of Proposition 8.8 verifies that g_A has the property that $(g_A(\mathbb{C})^{top})^{\hat{L}} \sim (g_{\bar{k}})_{et}^{\hat{L}}$ and $(g_A(\mathbb{C})^{top})^{\hat{L}'} \sim (g_{\bar{k}'})_{et}^{\hat{L}'}$, where L (respectively, L') denotes the set of all primes except the characteristic of k (resp., k'). This implies that $(g_A(\mathbb{C}))^{top}$ is homotopically trivial if $(g_{\bar{k}})_{et}$ and $(g_{\bar{k}'})_{et}$ are homotopically trivial.

Let $T' \to T$ be the finite, surjective, pointed covering space associated to the subgroup $(\ker \rho_1) \cap (\ker \rho_1')$ of $\pi_1(T,t)$, where $\rho_1: \pi_1(T,t) \to GL_n(k)$ and $\rho_1': \pi_1(T,t) \to GL_n(k')$ are induced by $\rho: \pi_1(T,t) \to GL_n(A)$. Applying the above argument to $\rho': \pi_1(T',t) \to GL_n(A)$, we conclude that the classifying map for $p': E' \to T'$ is homotopically trivial. ∎

The following result of D. Cox describes the etale homotopy type of a real algebraic variety in terms of its associated complex analytic variety.

PROPOSITION 11.2 (D. Cox [20]). *Let* X *be a connected real algebraic variety (i.e., a reduced, irreducible scheme of finite type over* $\operatorname{Spec}\mathbf{R}$ *), and let* $X_{\mathbf{C}}^{top}$ *denote the complex analytic space of complex points of* X. *Then there is a weak homotopy equivalence of Artin-Mazur completions (with respect to the set* P *of all primes)*

$$X_{et}^{\hat{P}} \sim (X_{\mathbf{C}}^{top} \underset{G}{\times} |EG|)^{\hat{P}}$$

where G *is the group* $\mathbf{Z}/2$ acting on $X_{\mathbf{C}}^{top}$ by complex conjugation.

Proof. Let $X_{\mathbf{C}} = X \underset{\operatorname{Spec}\mathbf{R}}{\times} \operatorname{Spec}\mathbf{C}$, so that $X_{\mathbf{C}} \to X$ is galois with group G. Let $U. = \operatorname{cosk}_0^X(X_{\mathbf{C}})$; because $U. \simeq X \underset{\operatorname{Spec}\mathbf{R}}{\times} \operatorname{cosk}_0^{\operatorname{Spec}\mathbf{R}}(\operatorname{Spec}\mathbf{C})$ and $\pi_0(\operatorname{cosk}_0^{\operatorname{Spec}\mathbf{R}}(\operatorname{Spec}\mathbf{C})) = BG$, there is a natural homeomorphism

$$|U.^{top}| \simeq X_{\mathbf{C}}^{top} \underset{G}{\times} |EG| .$$

The proposition now follows directly from Proposition 8.1 and Theorem 8.4. ■

The following criterion was first proved by M. Artin and J.-L. Verdier and published by D. Cox with a proof based on Proposition 11.2. We present another proof based on this proposition.

COROLLARY 11.3 (D. Cox [20]). *Let* X *be a connected real algebraic variety of dimension* n. *Then* X *has at least one real point if and only if* $H^i(X_{et}, \mathbf{Z}/2) \neq 0$ *for some* $i > 2n$.

Proof. If X has a real point, then this point is a section for the structure map $X \to \operatorname{Spec}\mathbf{R}$. Because $H_{et}^i(\operatorname{Spec}\mathbf{R}, \mathbf{Z}/2) = H^i(B\mathbf{Z}/2, \mathbf{Z}/2) \neq 0$ for any $i > 0$, we conclude that $H_{et}^i(X, \mathbf{Z}/2) \neq 0$ for any $i > 2n$ whenever X has a real point. Conversely, if X has no real point, then $G = \mathbf{Z}/2$ acts freely on $X_{\mathbf{C}}^{top}$, so that $X_{\mathbf{C}}^{top} \underset{G}{\times} |EG|$ is homotopy equivalent to $X_{\mathbf{C}}^{top}/G$. Because $X_{\mathbf{C}}^{top}/G$ is a complex analytic variety of complex dimension n and

$H^i(X_C^{top}/G, Z/2) \simeq H^i(X_{et}, Z/2)$ by Proposition 11.2, we conclude that $H^i(X_{et}, Z/2) = 0$ for all $i > 2n$ whenever X has no real points. ■

In [48], Corollary X.1.4, A. Grothendieck proved the exactness of the following sequence

$$\pi_1(\overline{X}_y) \to \pi_1(\overline{X}) \to \pi_1(Y) \to 1$$

whenever $\overline{f}: \overline{X} \to Y$ is a proper, smooth pointed map of noetherian, normal schemes with connected geometric fibre $\overline{X}_y$. This sequence was generalized by M. Raynaud based on notes of Grothendieck ([48], XIII.2.9 and XIII.4.1) to the assertion that the following sequence is exact

$$\pi_1(X_y)^{\hat{L}} \to \pi_1(X)/N \to \pi_1(Y) \to 1$$

where N is the normal subgroup of $\pi_1(X)$ defined by

$$N = \ker(\ker(\pi_1(X) \to \pi_1(Y)) \to \ker(\pi_1(X) \to \pi_1(Y))^{\hat{L}}) ,$$

L is the set of primes complementary to the residue characteristics of Y, and $f: X \to Y$ is the restriction of $\overline{f}: \overline{X} \to Y$ as above to $X = \overline{X} - D$ with D is a divisor in $\overline{X}$ with normal crossings over Y (as defined below).

We proceed to extend this latter homotopy sequence to higher homotopy groups, observing that L-completion is essential (for example, the Kunneth Theorem fails completely for coefficients not prime to residue characteristics). The homotopy sequence will be proved for "geometric fibrations" which we now define.

DEFINITION 11.4. A map $f: X \to Y$ of schemes is said to be a special geometric fibration if f is the restriction of a proper, smooth map $\overline{f}: \overline{X} \to Y$ to $\overline{X} - T$, where T is a closed subscheme of $\overline{X}$ satisfying the following condition: T is the union of closed subschemes T_i of pure codimension c_i in $\overline{X}$ over Y such that each non-empty intersection $T_{i_1} \cap \cdots \cap T_{i_s}$ is smooth over Y of pure codimension $c_{i_1} + \cdots + c_{i_s}$. (If each $c_i = 1$, then T is said to be a divisor in $\bar{X}$ with normal crossings over Y.) More generally, $f: X \to Y$ is said to be a *geometric fibration* if Y admits a Zariski open covering $\{V_j \to Y\}$ such that the restriction of f above each V_j, $f_| : f^{-1}(V_j) \to V_j$, is a special geometric fibration. A geometric fibration (of relative dimension 1) is said to be an *elementary fibration* if its geometric fibres are connected, affine curves. ∎

In the following theorem (Theorem 4.2 of [31]), our consideration of $\{f(i): X(i) \to Y(i)\}$ is a formal means of passing to the "*universal covering space*" of Y. This is necessary in order to compare the homotopy fibre of the L-completion to the L-completion of the homotopy fibre.

THEOREM 11.5. *Let* $f: X \to Y$ *be a pointed geometric fibration of noetherian, normal schemes with connected geometric fibre* X_y. *Let* $\{Y(i); i \in I\}$ *be the pro-object of pointed schemes consisting of an isomorphism class of each pointed galois covering* $Y(i) \to Y$ *of* Y, *and let* $f(i): X(i) = X \times_Y Y(i) \to Y(i)$ *be the pull-back of* f *by* $Y(i) \to Y$. *Then there exists a long exact homotopy sequence*

$$\cdots \to \pi_n((X_y)^{\hat{L}}_{et}) \to \pi_n(\{X(i)^{\hat{L}}_{et}\}) \to \pi_n(\{Y(i)^{\hat{L}}_{et}\}) \to \cdots$$

where L *is the set of primes complementary to the residue characteristics of* Y.

Proof (Sketch). By Lemma 10.6, the natural map $\text{fib}(\{f(i)_{et}\}) \to \text{fib}(f_{et})$ is a weak equivalence. Because $\{Y(i)_{et}\}$ is simply connected, the natural map $\text{fib}(\{f(i)_{et}\})^{\hat{L}} \to \text{fib}(\{f(i)^{\hat{L}}_{et}\})$ is a weak equivalence. Thus, it suffices to prove the natural map $(X_y)_{et} \to \text{fib}(f_{et})$ is a weak L-equivalence (i.e., satisfies conditions a) and b) of Corollary 6.5).

Let K denote $\ker(\pi_1(X_{et}) \to \pi_1(Y_{et}))^{\hat{L}}$. Because the removal of smooth closed subschemes of pure codimension greater than 1 does not affect the fundamental group ([7], XVI. 3.3), we may apply the above Raynaud exact sequence to conclude that $\pi_1((X_y)^{\hat{L}}_{et}) \to K$ is surjective with kernel H. One verifies that H is central in $\pi_1((X_y)^{\hat{L}}_{et})$. The homotopy sequence implies that H′, the kernel of the surjective map $\pi_1(\mathrm{fib}(f_{et})^{\hat{L}}) \to K$ is also central.

Let M be a locally constant, constructible, L-torsion abelian sheaf on X(i) whose restriction to some L-primary galois extension $X(i)' \to X(i)$ is constant. We verify below that we may apply Theorem 10.7 to f(i) and M, concluding that $(X_y)_{et} \to \mathrm{fib}(f_{et})$ induces isomorphisms

(11.5.1) $$H^*(\mathrm{fib}(f_{et}),N) \longrightarrow H^*((X_y)_{et},N)$$

for any local system N of finite, L-primary abelian groups on $\mathrm{fib}(f_{et})$ induced from a representation of K. To verify the hypotheses of Theorem 10.7 for f(i) and M, we may assume f(i) is a special geometric fibration with relative compactification $j: X(i) \to \overline{X}(i)$. As in the proof of the proper, smooth base change theorem ([7], XVI. 2.2), it suffices to prove that $R^q f(i)_* M$ commutes with arbitrary base change on Y(i) for all $q \geq 0$.

Using the Leray spectral sequence and the proper base change theorem for $\overline{f}(i)$, we are reduced to proving that $R^p j_* M$ commutes with arbitrary base change on Y(i) for all $p \geq 0$ in order to verify the hypotheses of Theorem 10.7 for f(i) and M. This is verified by examining the stalks of $R^p j_* M$ at a geometric point x of $\overline{X}(i) - X(i)$, computed as the limit of $H^p(\ \ ,M)$ applied to deleted etale neighborhoods of x. Finally, the effect upon $(R^q j_* M)_x$ of arbitrary base change on Y(i) is controlled using the cohomological purity theorem ([59], VI.5.1) thanks to the following consequence of Abhyankar's Lemma ([48], XII.5.5): on a cyclic covering of a deleted neighborhood of x, M can be extended to a locally constant sheaf on some smooth, relative (to Y(i)) compactification.

Let $((X_y)^{\hat{L}}_{et})_H \to (X_y)^{\hat{L}}_{et}$ and $(\mathrm{fib}(f_{et})^{\hat{L}})_{H'} \to \mathrm{fib}(f_{et})^{\hat{L}}$ be the covering spaces associated to H and H′. Then (11.5.1) implies that

(11.5.2) $$((X_y)^{\hat{L}}_{et})_H \to (\mathrm{fib}(f_{et})^{\hat{L}})_{H'}$$

induces an isomorphism in $\mathbf{Z}/\ell$ cohomology for each $\ell \in L$ and thus an isomorphism of (abelian) fundamental groups. The proof that (11.5.2) (and, thus, $(X_y)_{et}^{\hat{L}} \to \mathrm{fib}(f_{et})^{\hat{L}})$ is a weak equivalence is completed by proving that $H \simeq H'$ acts trivially on the $\mathbf{Z}/\ell$ cohomology of the universal covering spaces of $(X_y)_{et}^{\hat{L}}$ and $\mathrm{fib}(f_{et})^{\hat{L}}$. This last statement is proved using the following topological result ([31], Appendix): if $F \to E \to B$ is a fibre triple of connected, pointed spaces, then the action of $\pi_1(F)$ on the homology of the universal covering of F factors through an action of $\pi_1(E)$. ■

The geometric basis for our last application, Theorem 11.7, is the following theorem of M. Artin.

THEOREM 11.6 (M. Artin [7], XI. 3.3). *Let* X *be a smooth algebraic variety of dimension* n *over an algebraically closed field* k *and let* x *be a closed point of* X. *Then there exists a Zariski open neighborhood* U *of* x *whose structure map* $U \to \mathrm{Spec}\, k$ *may be factored as a composition of elementary fibrations*:

$$U = U_n \xrightarrow{f_n} U_{n-1} \longrightarrow \cdots \longrightarrow U_1 \xrightarrow{f_1} \mathrm{Spec}\, k \;. \;\blacksquare$$

In the special case in which k equals $\mathbf{C}$, Theorem 11.6 asserts that X^{top} admits the structure of the total space of an iterated fibration with fibres which are connected, noncompact Riemann surfaces. Because the latter are $K(\pi,1)$'s with π a free group, the homotopy sequence implies that X^{top} is also a $K(\pi,1)$ with π a successive extension of free groups.

To conclude a similar result for positive characteristic varieties, we must consider homotopy types completed away from the characteristic of k. The non-exactness of completion requires us to consider etale neighborhoods and only one prime at a time. The proof we provide for Theorem 10.7 is somewhat simpler than the original one given in [30].

THEOREM 11.7. *Let* X *be a smooth algebraic variety over an algebraically closed field* k, *let* x *be a closed point of* X, *and let* ℓ *be a prime invertible in* k. *Then there exists an etale neighborhood* V *of* x *in* X *such that* $\pi_1(V_{et})^{\hat{\ell}}$ *is a successive extension of finitely generated, free* pro-ℓ *groups and such that the natural map*

$$V_{et}^{\ddot{\ell}} \to K(\pi_1(V_{et})^{\ddot{\ell}},1)$$

is a weak equivalence.

Proof. The proof of Theorem 11.6 provided by M. Artin in [7] proceeds by verifying the existence of a Zariski open U_n of X together with an elementary fibration $f_n: U_n \to U_{n-1}$ where U_{n-1} is open in P^{n-1} (and therefore smooth). Let $g: U'_{n-1} \to U_{n-1}$ be a finite, etale map chosen so that $R^1 f'_{n*} Z/\ell \simeq g^* R^1 f_{n*} Z/\ell$ is constant on U'_{n-1}, where $f'_n : U'_n \to U'_{n-1}$ is the pull-back of f_n by g. If we iterate this procedure (next considering the smooth scheme U'_{n-1} in place of X), we obtain an etale neighborhood V of x in X whose structure map factors as a succession of elementary fibrations

$$V = V_n \xrightarrow{g_n} V_{n-1} \longrightarrow \cdots \longrightarrow V_1 \xrightarrow{g_1} \operatorname{Spec} k$$

with the property that $R^1 g_{i*} Z/\ell$ is constant on V_{i-1} for each i.

Let C_i denote the geometric fibre of g_i. As argued in the proof of Theorem 11.5, we may apply Theorem 10.7 to conclude that $(C_i)_{et} \to \operatorname{fib}((g_i)_{et})$ induces an isomorphism in Z/ℓ cohomology. Because C_i is connected and has ℓ-cohomological dimension 1, the action of $\pi_1((V_{i-1})_{et})$ on $H^*((C_i)_{et}, Z/\ell) \simeq H^*(\operatorname{fib}(g_i)_{et}, Z/\ell)$ is trivial (since $R^1 g_{i*} Z/\ell$ is constant). We may therefore compare the Serre spectral sequence for $(g_i)_{et}$ and $(g_i)_{et}^{\hat{\ell}}$ to conclude that $\operatorname{fib}((g_i)_{et}) \to \operatorname{fib}((g_i)_{et}^{\hat{\ell}})$ induces an isomorphism in Z/ℓ cohomology.

We conclude that the composition

$$(C_i)_{et} \to \operatorname{fib}((g_i)_{et}) \to \operatorname{fib}((g_i)_{et}^{\ddot{\ell}})$$

induces an isomorphism in $\mathbf{Z}/\ell$ cohomology and therefore a weak equivalence $(C_i)^{\hat{\ell}}_{et} \to \mathrm{fib}((g_i)^{\hat{\ell}}_{et})$ of ℓ-nilpotent pro-simplicial sets. Because C_i is a smooth, connected, affine curve over k, $(C_i)^{\hat{\ell}}_{et}$ is weakly equivalent to $K(\pi,1)$ where π is a finitely generated, free pro-ℓ group. The theorem now follows immediately by applying the long exact homotopy sequence to the successive fibering

$$V^{\hat{\ell}}_{et} \to (V_{n-1})^{\hat{\ell}}_{et} \to \cdots \to (V_1)^{\hat{\ell}}_{et} . \blacksquare$$

12. APPLICATIONS TO FINITE CHEVALLEY GROUPS

In this chapter, we employ etale homotopy theory to determine cohomology groups of finite Chevalley groups and homotopy types of associated K-theory spaces. In particular, our techniques apply to the group of F_q-rational points $G(F_q)$ of an algebraic group G_k over a field k of finite characteristic and to related twisted groups (for example, to

$$U_n(F_q) = \{(\alpha_{ij}) \in GL_n(\bar{F}_q), (\alpha_{ij}) \cdot (\alpha_{ij}^q)^t = I_n\}).$$

The basic theorem of this chapter is Theorem 12.2 which provides an etale homotopy theoretic interpretation of an isomorphism of S. Lang. Although Theorem 12.2 was first proved by the author in [34], the relevance of Lang's isomorphism was first observed by D. Quillen in [62]; Quillen also suggested formulating the Lang isomorphism as a homotopy cartesian square. Corollary 12.3 relates a stability result for the cohomology of finite classical groups, whereas Corollary 12.4 provides a comparison of the cohomology of discrete Chevalley groups to that of the classifying spaces of the associated Lie groups. Proposition 12.5 determines certain (unstable) spaces obtained using Quillen's plus construction. The most interesting application of Theorem 12.2 is given in Theorem 12.7, in which the various unitary K-theory spaces for finite fields are identified. As the reader can ascertain by comparing the material of this chapter to the author's various papers on these topics, our presentation here is somewhat simpler and more direct than that in the literature.

The following proposition is a generalization due to R. Steinberg [68] of a theorem of S. Lang [53] to *twisted Chevalley groups* $G_k(k)^{\phi}$.

PROPOSITION 12.1. *Let* G_k *be a connected linear algebraic group over an algebraically closed field and let* $\phi : G_k \to G_k$ *be a surjective endomorphism such that the group of* k*-rational points of* G_k *invariant under* ϕ, $H = G_k(k)^{\phi}$, *is finite. Then the "Lang map"*

$$1/\phi : G_k \to G_k$$

is a principal H*-fibration, where* $1/\phi$ *sends a* k*-rational point* g *to* $g \cdot \phi(g)^{-1}$. *Consequently,* $1/\phi$ *induces an isomorphism*

$$1/\phi : G_k/H \xrightarrow{\sim} G_k \,. \;\blacksquare$$

This proposition is particularly striking when one recalls that topological groups have no non-abelian connected covering spaces. Thus, Proposition 8.8 implies that whenever H (as in Proposition 12.1) is non-abelian and G reductive, the order of H must be divisible by the residue characteristic of k.

The following theorem (Theorem 2.9 of [34]) is our etale homotopy theoretic interpretation of Proposition 12.1. The square (12.2.1) has been referred to as the "*cohomological Lang fibre square.*"

THEOREM 12.2. *Let* G(C) *be a complex reductive Lie group, let* G_Z *be an associated Chevalley integral group scheme, let* k *be an algebraically closed field of characteristic* p, *and let* $\phi : G_k \to G_k$ *be a surjective endomorphism with* $H = G_k(k)^{\phi}$ *finite. Then a choice of embedding of the Witt vectors into* C *determines a commutative square in the homotopy category for any prime* ℓ *with* $p \nmid \ell$:

(12.2.1)
$$\begin{array}{ccc}
BH & \longrightarrow & (Z/\ell)_\infty \circ \mathrm{Sing.}(BG(C)) \\
\Big\downarrow D & & \Big\downarrow \Delta \\
(Z/\ell)_\infty \circ \mathrm{Sing.}(BG(C)) & \longrightarrow & (Z/\ell)_\infty \circ \mathrm{Sing.}(BG(C)^{\times 2})
\end{array}$$

with the property that a choice of map on homotopy fibres $\mathrm{fib}(D) \to \mathrm{fib}(\Delta)$ *determined by (12.2.1) induces isomorphisms in* $\mathbf{Z}/\ell$ *cohomology, where* Δ *is induced by the diagonal* $G \to G^{\times 2}$.

Proof. We interpret Proposition 12.1 as asserting that the following square of simplicial schemes

$$\begin{array}{ccc} B(G_k/H,G_k,*) & \longrightarrow & B(G_k^{\times 2}/\Delta(G_k),G_k^{\times 2},*) \\ \Big\downarrow d & & \Big\downarrow \delta \\ BG_k & \xrightarrow{1\times\phi} & BG_k^{\times 2} \end{array} \tag{12.2.2}$$

is cartesian, with the isomorphism on fibres $G_k/H \xrightarrow{\sim} G_k^{\times 2}/\Delta(G_k)$ given by the Lang isomorphism $G_k/H \xrightarrow{\sim} G_k$. The hypotheses of Theorem 10.7 are satisfied with $M = \mathbf{Z}/\ell$, because for each simplicial degree n both d_n and δ_n are product projections with smooth bases so that the smooth base change theorem implies that $R^q d_{n*}\mathbf{Z}/\ell$ and $R^q\delta_{n*}\mathbf{Z}/\ell$ are constant. Consequently, $\mathrm{fib}(d_{et}) \to \mathrm{fib}(\delta_{et})$ induces an isomorphism

$$H^*(\mathrm{fib}(\delta_{et}),\mathbf{Z}/\ell) \xrightarrow{\sim} H^*(\mathrm{fib}(d_{et}),\mathbf{Z}/\ell) ,$$

where $\mathrm{fib}(d_{et})$ and $\mathrm{fib}(\delta_{et})$ are defined in Definition 10.1.

Consequently, applying Proposition 8.8 to $(12.2.2)_{et}$ we obtain the commutative square in $\mathcal{H}_*$ inducing $H^*(\mathrm{fib}(\Delta),\mathbf{Z}/\ell) \xrightarrow{\sim} H^*(\mathrm{fib}(\ddot{d}_{et}),\mathbf{Z}/\ell)$

$$\begin{array}{ccc} B(G_k/H,G_k,*)_{et} & \longrightarrow & (\mathbf{Z}/\ell)_\infty \circ \mathrm{Sing.}(BG(\mathbf{C})) \\ \Big\downarrow \ddot{d}_{et} & & \Big\downarrow \Delta \\ (\mathbf{Z}/\ell)_\infty \circ \mathrm{Sing.}(BG(\mathbf{C})) & \longrightarrow & (\mathbf{Z}/\ell)_\infty \circ \mathrm{Sing.}(BG(\mathbf{C})) . \end{array}$$

To complete the proof of the theorem, we observe that

$$B(G_k,G_k,*) \to B(G_k/H,G_k,*)$$

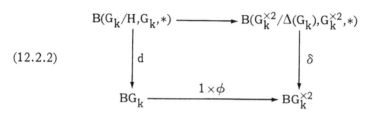

is a principal H-fibration so that $B(G_k/H,G_k,*)_{et}$ is weakly equivalent to BH provided that $B(G_k,G_k,*)_{et}$ is contractible (by Lemma 10.6). Yet a simplicial homotopy

$$B(G_k,G_k,*)\otimes\Delta[1] \to B(G_k,G_k,*)$$

is readily constructed (e.g., [34], Proposition 3.7) which relates the identity to the point map (factoring through the identity point $e: \operatorname{Spec} k \to B(G_k,G_k,*)$). Consequently, the contractibility of $B(G_k,G_k,*)_{et}$ follows from the contractibility of $(\operatorname{Spec} k)_{et}$ and Proposition 4.7. ∎

Theorem 12.2 provides a means of computing the Z/ℓ cohomology of the finite *(twisted) Chevalley group* $H = G_k(k)^\delta$ in terms of $H^*(BG,Z/\ell)$ for ℓ relatively prime to the residue characteristic of k. In his thesis [52], S. Kleinerman has employed the Eilenberg-Moore spectral sequence for (12.2.1) to determine $H^*(H,Z/\ell)$ in the case that H equals $G(F_q)$ for (exceptional) simple algebraic groups G such that $H_*(BG(C),Z)$ has no ℓ-torsion.

The following corollary is an immediate consequence of Theorem 12.2 and the corresponding *cohomological stability* for the cohomology of $BG_n(C)$. Corollary 12.3 first appeared as Theorem 2 of [35].

COROLLARY 12.3. *Let ℓ be an integer invertible in the finite field F_q. Then the following natural restrictions maps are isomorphisms:*

$$\begin{array}{rcll}
H^i(GL_{n+1}(F_q),Z/\ell) & \to & H^i(GL_n(F_q),Z/\ell) & i \le 2n \\
H^i(SL_{n+1}(F_q),Z/\ell) & \to & H^i(SL_n(F_q),Z/\ell) & i \le 2n \\
H^i(U_{n+1}(F_q),Z/\ell) & \to & H^i(U_n(F_q),Z/\ell) & i \le 2n \\
H^i(Sp_{2n+2}(F_q),Z/\ell) & \to & H^i(Sp_{2n}(F_q),Z/\ell) & i \le 4n+2 \\
H^i(SO_{n+1}(F_q),Z/\ell) & \to & H^i(SO_n(F_q),Z/\ell) & i \le n-2 \\
H^i(Spin_{n+1}(F_q),Z/\ell) & \to & H^i(Spin_n(F_q),Z/\ell) & i \le n-2 .
\end{array}$$ ∎

Another easy consequence of Theorem 12.2 is our next corollary which shows that BG(F) is a good cohomological model for BG(C). Corollary

12.3 includes Proposition 1.3 of [32] (all of the classical types) and Proposition 5.3 of [52] (each of the exceptional types with ℓ not dividing the order of the Weyl group).

COROLLARY 12.4. *Assume the notation of Theorem 12.2 and let* $F = \bar{F}_p$, *the algebraic closure of* F_p. *Assume that* G *is either of classical type or that*

$$H_*(T(C), Z/\ell) \to H_*(G(C), Z/\ell)$$

is surjective, where T *is a maximal torus of* G. *The direct limit (with respect to* $q = p^d$ *) of the maps*

$$D_q : BG(F_q) \to (Z/\ell)_\infty \circ \operatorname{Sing.}(BG(C))$$

associated to the frobenius maps $\phi^q : G_k \to G_k$ *induces isomorphisms in homology and cohomology*

$$H_*(BG(F), Z/\ell) \xrightarrow{\sim} H_*(BG(C), Z/\ell), \quad H^*(BG(C), Z/\ell) \xrightarrow{\sim} H^*(BG(F), Z/\ell).$$

Proof. Clearly, it suffices to consider homology. The map D_q arises from the projection $d_q : B(G_k/G(F_q), G_k, *) \to BG_k$, so that it suffices to show that these maps induce an isomorphism

$$\operatorname{colim} H_*(B(G_k/G(F_q), G_k, *), Z/\ell) \xrightarrow{\sim} H_*(BG_k, Z/\ell).$$

Using the colimit of the Serre spectral sequences for each $(d_q)_{et}$, we conclude that it suffices to prove that $\operatorname{colim} \tilde{H}_*(G_k/G(F_q), Z/\ell) = 0$.

To determine the map in homology induced by $G_k/G(F_q) \to G_k/G(F_{q'})$, we fit this map and the Lang isomorphisms into the following commutative square

$$\begin{array}{ccc} G_k/G(F_q) & \longrightarrow & G_k/G(F_{q'}) \\ \downarrow{\scriptstyle 1/\phi^q} & & \downarrow{\scriptstyle 1/\phi^{q'}} \\ G_k & \xrightarrow{\theta^{q'/q}} & G_k \end{array}$$

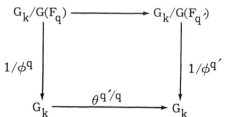

where $\theta^{q'/q}$ is the product of the maps $\phi^{q^i}: G_k \to G_k$ for $0 \le i < t$ with $q' = q^t$. We readily verify that the restriction of $\theta^{q'/q}$ to T_k induces $\theta_*^{q'/q}: H_1(T_k, Z/\ell) \to H_1(T_k, Z/\ell)$ given as multiplication by $1+q+\cdots+q^{t-1}$. Because $H^*(T_k, Z/\ell)$ is generated by $H^1(T_k, Z/\ell)$, we conclude that

$$\theta_*^{q'/q}: \tilde{H}_*(T_k, Z/\ell) \to \tilde{H}_*(T_k, Z/\ell)$$

is the 0-map whenever ℓ divides

$$(q'-1)/(q-1) = 1+q+\cdots+q^{t-1}.$$

If $H_*(T, Z/\ell) \to H_*(G, Z/\ell)$ is surjective, the naturality of Proposition 8.8 implies that

$$\theta_*^{q'/q}: \tilde{H}_*(G_k, Z/\ell) \to \tilde{H}_*(G_k, Z/\ell)$$

is also the 0-map for t sufficiently large, so that

$$\operatorname{colim} \tilde{H}_*(G_k/G_k(F_q), Z/\ell) = 0$$

as required. The remaining cases to consider are $G = SO_n$ or $Spin_n$, $\ell = 2$ and q odd. In this case,

$$\phi_*^{q^i}: H_*(G_k, Z/2) \to H_*(G_k, Z/2)$$

is seen to be the identity, because $\phi_*^{q^i}$ can be viewed as the restriction of

$$\Omega\Psi_*^{q^i}: H_*(SO, Z/2) \to H_*(SO, Z/2),\ \Omega\Psi_*^{q^i}: H_*(Spin, Z/2) \to H_*(Spin, Z/2)$$

(as in the proof of Theorem 12.7 below). Thus,

$$\theta_*: \tilde{H}_*(G_k, Z/2) \to \tilde{H}_*(G_k, Z/2)$$

is the 0-map for t even ($\theta^{q'/q*}$ is multiplication by t on primitives in $\tilde{H}^*(G_k, Z/2)$), so that $\operatorname{colim} \tilde{H}_*(G_k/G(F_q); Z/2) = 0$. ■

If G is a reductive group over an algebraically closed field of characteristic p and if $F = \overline{F}_p$, then the commutator subgroup of $G(F) = G_k(F)$ is perfect. We recall that the Quillen plus construction [43]

$$BG(F) \to BG(F)^+$$

(with respect to the commutator subgroup) induces an isomorphism in homology and is the abelianization map on fundamental groups.

Using a weight argument, one can easily verify that no cohomology class of $\tilde{H}^*(U(F),Z/p)$ is invariant under $T(F)$ for any reductive group G_k, where T_k is a maximal torus of G_k and U_k is the unipotent radical of a minimal parabolic. Because $U(F_q)$ contains a p-Sylow subgroup of $G(F_q)$ for any $q = p^d$, we conclude that $H^*(G(F),Z/p) = 0$ (cf. [64] or [34], Proposition 4.1).

This vanishing of the Z/p-cohomology of $G(F)$ and Corollary 12.4 enable us to determine the homotopy type of $BG(F)^+$ (as in [32], Theorem 2.2).

PROPOSITION 12.5. *Let* G *denote either* GL_n, SL_n, SO_n, $Spin_n$ *or* Sp_{2n} *for some* $n > 0$ *and let* $F = \overline{F}_p$ *for some prime* p. *Then the maps* $BG(F) \to (Z/\ell)_\infty \circ \mathrm{Sing.}(BG(C))$ *determine a map (unique up to homotopy on finite skeleta)*

$$BG(F)^+ \to (Z/\ell)_\infty \circ \mathrm{Sing.}(BG(C))$$

which can be identified with the fibre $Q/Z_{(p)}(BG(C))$ *of the map*

$$\mathrm{Sing.}(BG) \to (Z_{(p)})_\infty \circ \mathrm{Sing.}(BG)\ ,$$

where $Z_{(p)}$ *is the subring of* Q *consisting of rationals whose denominators are not divisible by* p.

Proof. We employ D. Sullivan's "arithmetic fibre square" technique to conclude that the vanishing of $\tilde{H}^*(BG(F),Z/p)$ and $\tilde{H}^*(BG(F),Q)$ imply that the maps

$$BG(F) \to (Z/\ell)_\infty \circ \mathrm{Sing.}(BG(C))$$

determine a map $BG(F) \to Sing.BG(C)$. The uniqueness up to homotopy of this map when restricted to finite skeleta of $BG(F)$ is given by ([13], VI. 8.1). Because $BG(C)$ is simply connected, this map factors uniquely through a map $BG(F)^+ \to Sing.(BG(C))$; because $BG(F)$ has trivial Z/p and Q homology, this latter map uniquely factors through a map

$$BG(F)^+ \to (Q/Z_{(p)})(BG(C)) .$$

By Corollary 12.4, this map induces an isomorphism in integral homology; because $\pi_1(BG(F)^+)$ is abelian, we conclude that this map induces an isomorphism on fundamental groups. For $G = SL_n$, $Spin_n$, or Sp_{2n}, $BG(C)$ is 2-connected so that $(Q/Z_{(p)})(BG(C))$ is simply connected; the Whitehead theorem then implies that $BG(F)^+ \to (Q/Z_{(p)})(BG(C))$ is a homotopy equivalence. Because

$$\pi_2(Sing.(BSO_n(C))) \to \pi_2((Z_{(p)})_\infty \circ Sing.(BSO_n(C)))$$

is surjective for all p, we similarly conclude that $BSO_n(F) \to (Q/Z_{(p)})(BSO_n(C))$ is a homotopy equivalence. Finally, $BGL_n(F)^+ \to (Q/Z_{(p)})(BGL_n(C))$ is a homotopy equivalence, because the induced map on universal covering spaces is $BSL_n(F)^+ \to (Q/Z_{(p)})(BSL_n(C))$. ■

In order to state the most interesting consequence of Theorem 12.2, we must recall the following definition (Definition 1.2 of [34]) of the *classical Chevalley groups*.

DEFINITION 12.6. Let q be a prime power. Then $F\Psi^q_G$ denotes the singular complex of the homotopy theoretic fibre of the map

$$d(1,\Psi^q_G) : BG(C) \to BG(C)$$

in the following four cases, given with discrete group $G(F_q)$

(i) $G(F_q)$ equals $GL(F_q)$ and $d(1,\Psi^q_{GL}) : BGL(C) \to BGL(C)$ represents $1-\Psi^q$ on complex K-theory.

(ii) $G(F_q)$ equals $SO(F_q)$ and $d(1,\Psi^q_{SO}): BSO(C) \to BSO(C)$ represents $1-\Psi^q$ on oriented real K-theory.

(iii) $G(F_q)$ equals $Sp(F_q)$ and $d(1,\Psi^q_{Sp}): BSp(C) \to BSp(C)$ represents $1-\Psi^q$ on symplectic K-theory with q odd.

(iv) $G(F_q)$ equals $U(F_q)$ and $d(1,\Psi^q_U): BGL(C) \to BGL(C)$ represents $1-\Psi^{-q}$ on complex K-theory. ∎

The following theorem, the fundamental result of [34], easily determines the *K-theories of finite fields* (the homotopy groups of $BGL(F_q)^+$, $BSO(F_q)^+$, $BSp(F_q)^+$, and $BU(F_q)^+$). These "unitary" K-groups of the finite field F_q have been tabulated in Theorem 1.7 of [34]. The original computation of $\pi_*(BGL(F_q)^+)$ was achieved by D. Quillen in [64] by other methods.

THEOREM 12.7. *With the notation of Definition 12.7, there are homotopy equivalences associated to the squares (12.2.1)*

$$\chi : BG(F_q)^+ \to F\Psi^q_G$$

for each of the four cases $G = GL$, SO, Sp, *and* U *of Definition 12.1.*

Proof. One verifies that $\tilde{H}^*(BG(F_q), Z/p) = 0$ where q is a power of the prime p by using a vanishing range for $\tilde{H}^*(BG(F_{q^d}), Z/p)$ which increases as d increases and by employing the transfer

$$H^*(BG(F_q), Z/p) \to H^*(BG(F_{q^d}), Z/p)$$

([34], Theorem 1.4). Because both $BG(F_q)^+$ and $F\Psi^q_G$ are simple spaces (in fact, infinite loop spaces) with trivial cohomology for Z/p and Q coefficients, it suffices to exhibit χ such that

$$\chi^* : H^*(BG(F_q)^+, Z/\ell) \to H^*(F\Psi^q_G, Z/\ell)$$

is an isomorphism for all primes $\ell \neq p$. As in the proof of Proposition 12.5, we conclude that it suffices to exhibit maps

$$\chi_\ell : BG(F_q) \to (Z/\ell)_\infty \circ F\Psi_G^q$$

for each prime $\ell \neq p$ inducing isomorphisms

$$\chi_\ell^* : H^*((Z/\ell)_\infty \circ F\Psi_G^q, Z/\ell) \to H^*(BG(F_q), Z/\ell) .$$

The colimit of (12.2.1) with respect to n for $G_k = G_{n,k}$ equal to $GL_{n,k}$, $SO_{n,k}$, $Sp_{2n,k}$, or $GL_{n,k}$ and ϕ equal to ϕ^q, ϕ^q, ϕ^q, or $(\ )^t \circ (\ )^{-1} \circ \phi^q$ determines a commutative square

(12.7.1)
$$\begin{array}{ccc} BG(F_q) & \longrightarrow & (Z/\ell)_\infty \circ \mathrm{Sing.}(BG(C)) \\ \downarrow & & \downarrow \Delta \\ (Z/\ell)_\infty \circ \mathrm{Sing.}(BG(C)) & \longrightarrow & (Z/\ell)_\infty \circ \mathrm{Sing.}(BG(C)^{\times 2}) . \end{array}$$

As is shown in Proposition 2.11 of [34], the lower horizontal arrow of (12.7.1) is $1 \times \Psi_G^q$, since $\phi^q : BG_{n,k} \to BG_{n,k}$ and the homotopy equivalences of Proposition 8.8 determine maps

$$(Z/\ell)_\infty \circ \mathrm{Sing.}(BG_n(C)) \to (Z/\ell)_\infty \circ \mathrm{Sing.}(BG_n(C))$$

which stabilize to $\Psi^q : BG(C) \to BG(C)$ (so that

$$(\ )^t \circ (\ )^{-1} \circ \phi^q : BGL_{n,k} \to BGL_{n,k}$$

determines $\Psi^{-q} : BGL(C) \to BGL(C)$).

By definition of $F\Psi_G^q$, $F\Psi_G^q$ fits in a homotopy cartesian square

(12.7.2)
$$\begin{array}{ccc} F\Psi_G^q & \longrightarrow & \mathrm{Sing.}(BG(C)) \\ \downarrow & & \downarrow \Delta \\ \mathrm{Sing.}(BG(C)) & \xrightarrow{1 \times \Psi_G^q} & \mathrm{Sing.}(BG(C))^{\times 2} \end{array}$$

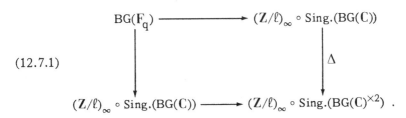

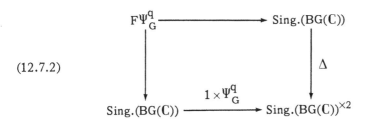

Because $BG(\mathbf{C})$ and $BG(\mathbf{C})^{\times 2}$ are simply connected, $(\mathbf{Z}/\ell)_\infty$ applied to (12.7.2) yields another homotopy cartesian square $(\mathbf{Z}/\ell)_\infty \circ (12.7.2)$. Therefore, we may choose a map between squares (in the homotopy category)

$$\zeta : (12.7.1) \to (\mathbf{Z}/\ell)_\infty \circ (12.7.2)$$

with the map on upper left corner defined to be χ_ℓ and the map on the other corners taken to be the identity. The fact that

$$\chi_\ell^* : H^*((Z/\ell)_\infty \circ F\Psi^q_G, Z/\ell) \to H^*(BG(F_q), Z/\ell)$$

is an isomorphism follows from the observation that ζ induces an isomorphism of Eilenberg-Moore spectral sequences (or that ζ restricted to left vertical arrows induces an isomorphism of Serre spectral sequences). ∎

An application of Theorem 12.7 proved by the author and S. Priddy is given in [41]. The reader is referred to that paper for details.

13. FUNCTION COMPLEXES

Recent work by the author [39] and by the author and W. Dwyer [27] relating algebraic K-theory to topological K-theory has required the use of function complexes of etale topological types. This chapter is intended to provide the foundational material necessary for those applications as well as those envisioned in the future.

Much of this chapter (Propositions 13.4, 13.6, and Corollary 13.7) is devoted to proving that maps of domain or range satisfying certain properties induce homotopy equivalences of function complexes. These results enable one to use the finiteness theorems of Chapter 7 and the comparison theorems of Chapter 8 to partially identify various function complexes. The specific situation relevant to algebraic K-theory is treated in Proposition 13.10, based on finiteness properties verified in Corollary 13.9. The relationship of these function complexes to algebraic structures is described in Proposition 13.2.

We recall that the *function complex* $\mathrm{Hom}(S.\,,T.)$ associated to pointed simplicial sets $S.$ and $T.$ (with base points s and t respectively) is the pointed simplicial set whose set of n-simplices is the set of maps of pairs of simplicial sets

$$\mathrm{Hom}_n(S.\,,T) = \mathrm{Hom}((S. \times \Delta[n], \{s\} \times \Delta[n]), (T.\,,t)) .$$

For all pointed simplicial sets $R.\,,\ S.\,,\ T.\,,$ there is a canonical isomorphism

$$\mathrm{Hom}.\,(R.\,,\mathrm{Hom}.\,(S.\,,T.)) = \mathrm{Hom}.\,(R. \wedge S.\,,T.)$$

where $R. \wedge S. = R. \times S.\,/(\{r\} \times S. \cup R. \times \{s\})$. If $T.$ is a Kan complex, then $\mathrm{Hom}.\,(S.\,,T.)$ is also a Kan complex so that the n-th homotopy group of the component of $\mathrm{Hom}.(S.,T.)$ containing $f: S. \to T.$ consists of homotopy classes of maps $F: S. \times \Delta[n] \to T.$ relative to $f \circ pr_1 \cup t: S. \times sk_{k-1}\Delta[n] \cup \{s\} \times \Delta[n] \to T.$.

We extend $\mathrm{Hom}.(\ ,\ )$ to a functor

$$\mathrm{Hom}.(\ ,\ ): \text{pro-(s. sets}_*) \times \text{pro-(s. sets}_*) \to \text{pro-(s. sets}_*)$$

by defining

$$(13.1) \quad \mathrm{Hom}.(\{S.^i\,;i \in I\},\{T.^j\,;j \in J\}) = \{\operatorname*{colim}_I \{\mathrm{Hom}.(S.^i,T.^j);\, i \in I\};\, j \in J\}.$$

In the following proposition, we introduce the algebraic function complex and relate it to the function complex of etale topological types.

PROPOSITION 13.2. *For pointed simplicial schemes* $X.,x$ *and* $Y.,y$, *we define the algebraic function complex*

$$\mathrm{Hom}.(X.,Y.) \in (\text{s. sets}_*)$$

(with base points implicit) to be the simplicial set whose n*-simplices are maps of pairs* $(X. \otimes \Delta[n], \{x\} \otimes \Delta[n]) \to (Y.,y)$. *If* $X.$ *and* $Y.$ *are locally noetherian, then there is a natural map of pointed simplicial sets*

$$\mathrm{Hom}.(X.,Y.) \to \varprojlim_{HRR(Y.)} \mathrm{Hom}.((X.)_{et},(Y.)_{et}).$$

Proof. This map is obtained by observing that an n-simplex of $\mathrm{Hom}.(X.,Y.)$ represented by $X. \otimes \Delta[n] \to Y.$ determines $(X. \otimes \Delta[n])_{et} \to (Y.)_{et}$ in $\varprojlim \mathrm{Hom}_0((X. \otimes \Delta[n])_{et},(Y.)_{et})$ and thus in

$$\varprojlim \mathrm{Hom}_0((X.)_{et} \times \Delta[n] \times \Delta[n]/\{x\} \times \Delta[n],(Y.)_{et}) \simeq \varprojlim \mathrm{Hom}_n((X.)_{et},(Y.)_{et})$$

by Proposition 4.7. ∎

As an immediate corollary of Proposition 13.2, we conclude the existence of a natural map

(13.3) $$\text{Hom.}(X.\,,Y.) \to \underset{\substack{\leftarrow \\ HRR(Y.)}}{\text{holim}}\ \text{Hom.}((X.)_{et},(Y.)_{et})$$

thanks to the natural transformation $\underset{\leftarrow}{\lim}(\ ) \to \underset{\leftarrow}{\text{holim}}(\ )$ of [12], XI. 3.5. Because $\underset{\leftarrow}{\text{holim}}(\ )$ is functorial with respect to strict maps in pro-(s. sets$_*$), (13.2) is natural with respect to both X. and Y. .

We observe that Hom. $(\{S^i_.\},\ )$ determines a functor on the homotopy category $\mathcal{H}_*$: a pointed homotopy $H:(T.\times\Delta[1],\{t\}\times\Delta[1])\to(W.\,,w)$ between maps $f,g: T.\to W.$ of pointed Kan complexes induces a pointed homotopy between the maps $f_*,g_*:\text{Hom.}(\{S^i_.\},T.)\to\text{Hom.}(\{S^i_.\},W.)$ (of pointed Kan complexes). Consequently, we immediately conclude the following.

PROPOSITION 13.4. *Let* $\{S^i_.\}\in$ pro-(s. sets$_*$). *Then* Hom. $(\{S^i_.\},\ )$ *determines a functor*

$$\text{Hom.}(\{S^i_.\},\ ): \text{pro-}\mathcal{H}_* \to \text{pro-}\mathcal{H}_*\,.$$

In particular, if $\{T^j_.\}\to\{W^k_.\}$ *is a map in* pro-(Kan$_*$) (*where* (Kan$_*$) *is the full subcategory of* (s. sets$_*$) *consisting of Kan complexes*) *which is an isomorphism in* pro-$\mathcal{H}_*$, *then*

$$\text{Hom.}(\{S^i_.\},\{T^j_.\}) \to \text{Hom.}(\{S^i_.\},\{W^k_.\})$$

is also a map in pro-(Kan$_*$) *which is an isomorphism in* pro-$\mathcal{H}_*$. ∎

Let S. be a pointed simplicial set and T. a pointed Kan complex. For any $m\geq 0$, there is a natural isomorphism $\pi_m(\text{Hom.}(S.\,,T.))\simeq \text{Hom}_{\mathcal{H}_*}(\Sigma^m S.\,,T.)$, where the homotopy groups of Hom. (S. ,T.) are those based at the point map S. → T. and where $\Sigma^m S. = \Delta[m]\times S./(sk_{m-1}\Delta[m]\times S.\cup\Delta[m]\times\{s\})$ for $m>0$ and $\Sigma^0 S. = S.$. More generally, for $\{S^i_.\,;i\in I\}\in$ pro-(s. sets$_*$) and T. a pointed Kan complex, there are natural isomorphisms

$$\pi_m(\text{Hom.}(\{S^i_.\},T.)) \simeq \underset{I}{\text{colim}}\,\pi_m(\text{Hom.}(S^i_.\,,T.)) \simeq \text{Hom}_{\text{pro-}\mathcal{H}_*}(\{\Sigma^m S^i_.\},T.)$$

for any $m \geq 0$. Consequently, for any $\{S_.^i\} \in$ pro-(s. sets$_*$) and any $\{T_.^j ; j \in J\} \in$ pro-$\mathcal{H}_*$, we conclude the following natural isomorphism for any $m \geq 0$

$$\text{(13.5)} \qquad \pi_m(\mathrm{Hom.}(\{S_.^i\}, \{T_.^j\}) \simeq \{\mathrm{Hom}_{\text{pro-}\mathcal{H}_*}(\{\Sigma^m S_.^i\}, T_.^j); j \in J\}.$$

In the following proposition, we employ *obstruction theory* to study the behavior of Hom. (,) with respect to the first variable.

PROPOSITION 13.6. *Let* T. *be a pointed, connected Kan complex and let* $f : \{R_.^k : k \in K\} \to \{S_.^i : i \in I\}$ *be a map in* pro-(s. sets$_{*c}$). *Assume the following*: a.) $f^* : \mathrm{Hom}(\pi_1(\{S_.^i\}), \pi_1(T.)) \to \mathrm{Hom}(\pi_1(\{R_.^k\}), \pi_1(T.))$ *is a bijection;* b.) *for every map* $g : \pi_1(\{S_.^i\}) \to \pi_1(T.)$ *and every* $n > 1$, $f^* : H^m(\{S_.^i\}, g^*(\underset{\sim}{\pi}_n(T.))) \to H^m(\{R_.^k\}, f^* \circ g^*(\underset{\sim}{\pi}_n(T.)))$ *is an isomorphism for* $m \leq n+1$ (*where* $\underset{\sim}{\pi}_n(T.)$ *is the abelian local coefficient system on* T. *associated to the action of* $\pi_1(T.)$ *on* $\pi_n(T.)$); *and* c.) *there exists an integer* N *such that* $\pi_m(T.) = 0$ *for* $m \geq N$. *Then* f *induces a homotopy equivalence*

$$f^* : \mathrm{Hom.}(\{S_.^i\}, T.) \to \mathrm{Hom.}(\{R_.^k\}, T.)\,.$$

Proof. Because Hom. $(\{S_.^i\}, T.)$ and Hom. $(\{R_.^k\}, T.)$ are Kan complexes, it suffices to prove that f^* induces a bijection on connected components and an isomorphism on homotopy groups of corresponding connected components. Let $g, h : \{S_.^i\} \to T.$ be maps such that $g \circ f, h \circ f : \{R_.^k\} \to T.$ are homotopic. We inductively prove that for all $n \geq 2$ the compositions $g_{(n)}, h_{(n)} : \{S_.^i\} \to T. \to \mathrm{cosk}_n\, T.$ are homotopic; by c.), this will imply that g (homotopic to $g_{(N)}$) is homotopic to h (homotopic to $h_{(N)}$). Because $\mathrm{cosk}_2 T.$ is homotopy equivalent to $K(\pi_1(T.), 1)$, a.) implies that $g_{(2)}$ is homotopic to $h_{(2)}$. If $g_{(n)}$ and $h_{(n)}$ are homotopic, then choose i such that $g_{(n+1)}, h_{(n+1)}$ are represented by maps $S_.^i \rightrightarrows \mathrm{cosk}_{n+1}\, T.$. Choose $F : S_.^i \times \Delta[1] \to \mathrm{cosk}_n\, T.$ relating $g_{(n)}$, $h_{(n)}$ and $\alpha : R_.^k \to S_.^i$ representing $pr_i \circ f$ with $G : R_.^k \times \Delta[1] \to \mathrm{cosk}_{n+1}\, T.$ relating $g_{(n+1)} \circ \alpha$, $h_{(n+1)} \circ \alpha$ and lifting $F \circ (\alpha \times 1)$. We consider

$$\begin{array}{ccccc} R^k_\cdot \times\{0,1\}\cup\{r\}\times\Delta[1] & \longrightarrow & S^i_\cdot\times\{0,1\}\cup\{s\}\times\Delta[1] & \xrightarrow{g_{(n+1)},h_{(n+1)}} & \operatorname{cosk}_{n+1}T. \\ \downarrow & & \downarrow & & \downarrow \\ R^k_\cdot\times\Delta[1] & \longrightarrow & S^i_\cdot\times\Delta[1] & \xrightarrow{g_{(n)},h_{(n)}} & \operatorname{cosk}_n T.\,. \end{array}$$

Because the obstruction to finding a map $S^i_\cdot \times \Delta[1] \to \operatorname{cosk}_{n+1} T.$ which is a lifting up to homotopy for the right hand square lies in $H^n(S^i_\cdot, g^*(\underset{\sim}{\pi}_n(T.))$ and because the outer square admits such a lifting $R^k_\cdot \times \Delta[1] \to \operatorname{cosk}_{n+1} T.$, b.) implies that for some $i' \to i$ the compositions (also representing $g_{(n+1)}, h_{(n+1)}$) $S^{i'}_\cdot \rightrightarrows S^i_\cdot \to \operatorname{cosk}_{n+1} T.$ are homotopic. Thus, f^* induces an injection $\pi_0(\operatorname{Hom.}(\{S^i_\cdot\}, T.)) \to \pi_0(\operatorname{Hom.}(\{R^k_\cdot\}, T.))$ by (13.5).

To prove surjectivity on connected components, let $g' : \{R^k_\cdot\} \to T.$ be given. Then $g'_{(2)}$ extends (up to homotopy) to $g_{(2)} : \{S^i_\cdot\} \to \operatorname{cosk}_2 T.$ by a.). Proceeding inductively, let $S^i_\cdot \to \operatorname{cosk}_n T.$ represent $g_{(n)}$ extending $g'_{(n)}$ up to homotopy and let $R^k \to S^i_\cdot$ be such that $R^k_\cdot \to S^i_\cdot \to \operatorname{cosk}_n T.$ lifts to $R^k_\cdot \to \operatorname{cosk}_{n+1} T.$. To find $g_{(n+1)}$ extending $g'_{(n+1)}$, we must find $i' \to i$ such that there exists a map $S^{i'}_\cdot \to \operatorname{cosk}_{n+1} T.$ which is a lifting up to homotopy for the following diagram for some $R^{k'} \to S^{i'}_\cdot$ representing $pr_{i'} \circ f$:

$$\begin{array}{ccccc} R^{k'}_\cdot & \longrightarrow & R^k_\cdot & \longrightarrow & \operatorname{cosk}_{n+1} T. \\ \downarrow & & \downarrow & & \downarrow \\ S^{i'}_\cdot & \longrightarrow & S^i_\cdot & \longrightarrow & \operatorname{cosk}_n T. \end{array}$$

The existence of such a lifting is implied by the injectivity of $H^{n+1}(\{S^i_\cdot\}, g^*(\underset{\sim}{\pi}_n(T.))) \to H^{n+1}(\{R^k_\cdot\}, f^* \circ g^*(\underset{\sim}{\pi}_n(T.)))$ and the surjectivity of $H^n(\{S^i_\cdot\}, g^*(\underset{\sim}{\pi}_n(T.))) \to H^n(\{R^k_\cdot\}, f^* \circ g^*(\underset{\sim}{\pi}_n(T.)))$ (i.e., the vanishing of the (n+1)-st relative cohomology group) given by b.). Surjectivity is now implied by c.).

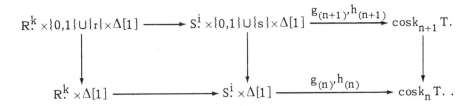

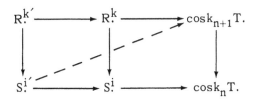

We now fix a base point $g:\{S^i_\cdot\}\to T_\cdot$ of some connected component of $\mathrm{Hom}_\cdot(\{S^i_\cdot\},T_\cdot)$. To prove that f^* induces an injection $\pi_m(\mathrm{Hom}_\cdot(\{S^i_\cdot\},T_\cdot),g) \to \pi_m(\mathrm{Hom}_\cdot(\{R^k_\cdot\},T_\cdot),g\circ f)$, it suffices to obtain liftings up to homotopy of diagrams of the form

$$\begin{array}{ccccc}
(R^k_\cdot\times sk_m(\Delta[m]\times\Delta[1]))\cup(\{r\}\times\Delta[m]\times\Delta[1]) & \rightarrowtail & (S^i_\cdot\times sk_m(\Delta[m]\times\Delta[1]))\cup(\{s\}\times\Delta[m]\times\Delta[1]) & \rightarrowtail & \mathrm{cosk}_{n+1}T_\cdot \\
\downarrow & & \downarrow & \nearrow & \downarrow \\
R^k_\cdot\times\Delta[m]\times\Delta[1] & \longrightarrow & S^i_\cdot\times\Delta[m]\times\Delta[1] & \longrightarrow & \mathrm{cosk}_n T_\cdot
\end{array}$$

provided that the outer square admits a lifting up to homotopy (after restricting to some $S^{i'}_\cdot\to S^i_\cdot$). To prove that f^* induces a surjection on m-th homotopy groups, it suffices to obtain liftings up to homotopy of diagrams of the form

$$\begin{array}{ccccc}
(R^k_\cdot\times sk_{m-1}\Delta[m])\cup(\{r\}\times\Delta[m]) & \rightarrowtail & (S^i_\cdot\times sk_{m-1}\Delta[m])\cup(\{s\}\times\Delta[m]) & \rightarrowtail & \mathrm{cosk}_{n+1}T_\cdot \\
\downarrow & & \downarrow & \nearrow & \downarrow \\
R^k_\cdot\times\Delta[m] & \longrightarrow & S^i_\cdot\times\Delta[m] & \longrightarrow & \mathrm{cosk}_n T_\cdot
\end{array}$$

(with a further arrow $R^k_\cdot\times\Delta[m]\to\mathrm{cosk}_{n+1}T_\cdot$)

(after restricting to some $R^{k'}_\cdot\to S^{i'}_\cdot$). These liftings are obtained as above using b.). ∎

As an immediate corollary of Proposition 13.6, we conclude the following.

COROLLARY 13.7. *Let* $\{T^j_\cdot\,;\,j\in J\}\in\text{pro-}(\mathrm{Kan}_{*c})$ *be such that each* $T^j_\cdot$ *has only finitely many non-zero homotopy groups and let* $f:\{R^k_\cdot\}\to\{S^i_\cdot\}$ *be a map in* pro-(s. sets$_{*c}$). *If* f *is a weak equivalence in* pro-$\mathcal{H}_*$, *then* $f^*:\mathrm{Hom}_\cdot(\{S^i_\cdot\},\{T^j_\cdot\})\to\mathrm{Hom}_\cdot(\{R^k_\cdot\},\{T^j_\cdot\})$ *in* pro-(Kan$_*$) *is an isomorphism in* pro-$\mathcal{H}_*$. *If* L *is a set of primes such that* $\pi_n(T^j_\cdot)$ *is finite* L*-torsion for each* $n\geq 1$, $j\in J$, *and if* $\hat{f}^L$ *is a weak equivalence in* pro-$\mathcal{H}_*$, *then* $f^*:\mathrm{Hom}_\cdot(\{S^i_\cdot\},\{T^j_\cdot\})\to\mathrm{Hom}_\cdot(\{R^k_\cdot\},\{T^j_\cdot\})$ *is an isomorphism in* pro-$\mathcal{H}_*$. ∎

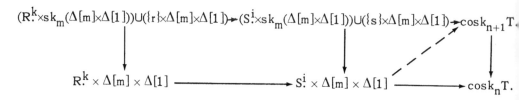

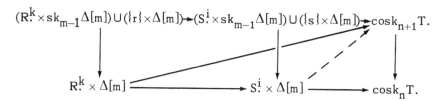

Employing the same techniques of obstruction theory as in the proof of Proposition 13.6, we obtain the following useful finiteness property.

PROPOSITION 13.8. *Let* $T.$ *be a pointed, connected Kan complex with finitely many non-zero homotopy groups and let* $\{S^i_\cdot\} \in$ pro-(s. sets$_{*c}$). *Assume that the set* $\mathrm{Hom}(\pi_1(\{S^i_\cdot\}), \pi_1(T.))$ *is finite and that for each* $g: \pi_1(\{S^i_\cdot\}) \to \pi_1(T.)$, *each* $n > 1$, *and each* $m \leq n$, $H^m(\{S^i_\cdot\}, g^*(\underset{\sim}{\pi}_n(T.)))$ *is finite. Then* $\mathrm{Hom.}(\{S^i_\cdot\},T.)$ *has finitely many components each of which has finite homotopy groups.*

Proof. As argued in the proof of Proposition 13.6, obstruction theory bounds the order of $\pi_0(\mathrm{Hom.}(\{S^i_\cdot\},T.))$ by the sum of the orders of the groups $\prod_{n>1} H^n(\{S^i_\cdot\}, g^*(\underset{\sim}{\pi}_n(T.)))$ indexed by $g \in \mathrm{Hom}(\pi_1(\{S^i_\cdot\}), \pi_1(T.))$. If $g: \{S^i_\cdot\} \to T.$ serves as a base point for some component of $\mathrm{Hom.}(\{S^i_\cdot\},T.)$, then obstruction theory once again bounds the order of $\pi_m(\mathrm{Hom.}(\{S^i_\cdot\},T.),g)$ by the order of $\prod_{n \geq m} H^{n-m}(\{S^i_\cdot\}, g^*(\underset{\sim}{\pi}_n(T.)))$ for any $m > 0$. ∎

As a corollary of Proposition 13.8 in conjunction with Corollary 6.10 and (13.5), we obtain the following determination of the homotopy groups of $\underleftarrow{\mathrm{holim}}\ \mathrm{Hom.}(\{S^i_\cdot\},\{T^j_\cdot\})$. The corollary is stated so as to be applicable to $\# \circ (Z/\ell)_\infty(BG_k)_{et}$ isomorphic in pro-$\mathcal{H}_*$ to $\# \circ (Z/\ell)_\infty(BG_k)_{nht} = \{W^k_\cdot ; k \in K\}$, where the homotopy groups of each $W^k_\cdot$ are finite by Theorem 7.3.

COROLLARY 13.9. *Let* $\{T^j_\cdot ; j \in J\} \in$ pro-(Kan$_{*c}$) *be isomorphic to* $\{W^k_\cdot ; k \in K\} \in$ pro-$\dot{\mathcal{H}}_{*c}$ *with each* $W^k_\cdot$ *having finitely many non-zero homotopy groups. Let* $\{S^i_\cdot\} \in$ pro-(s. sets$_{*c}$) *be such that for each* $k \in K$ *the set* $\mathrm{Hom}(\pi_1(\{S^i_\cdot\}), \pi_1(W^k_\cdot))$ *is finite and for each* $g: \pi_1(\{S^i_\cdot\}) \to \pi_1(W^k_\cdot)$ *the group* $\prod_{m \leq n} H^m(\{S^i_\cdot\}, g^*(\underset{\sim}{\pi}_n(W^k_\cdot)))$ *is also finite. Then for any* $n > 0$, *the homotopy groups of each component of* $\mathrm{Hom.}(\{S^i_\cdot\},\{T^j_\cdot\})$ *are pro-finite pro-groups. In particular, there are natural isomorphisms*

$$\pi_n(\underset{\leftarrow}{\text{holim}}\ \text{Hom.}(\{S^i_.\},\{T^j_.\})) \simeq \underset{\leftarrow}{\lim}\ \pi_n(\text{Hom.}(\{S^i_.\},\{T^j_.\}))$$

$$\simeq \text{Hom}_{\text{pro-}\mathcal{H}_*}(\{\Sigma^n S^i_.\},\{T^j_.\})\ .$$

Proof. By Proposition 13.8, the homotopy groups of $\text{Hom.}(\{S^i_.\},W^k_.)$ are finite for each $k \in K$. By Proposition 13.4, $\text{Hom.}(\{S^i_.\},\{T^j_.\})$ is isomorphic to $\text{Hom.}(\{S^i_.\},\{W^k_.\})$ in pro-$\mathcal{H}_*$, so that the homotopy pro-groups of $\text{Hom.}(\{S^i_.\},\{T^j_.\})$ are pro-finite. By Proposition 6.10, this implies the first asserted isomorphism. The second isomorphism follows immediately from (13.5). ■

The reader should not confuse the homotopy groups of $\underset{\leftarrow}{\text{holim}}\ \text{Hom.}(\{S^i_.\},\{T^j_.\})$ with those of $\text{Hom.}(\underset{\leftarrow}{\text{holim}}\{S^i_.\},\underset{\leftarrow}{\text{holim}}\{T^j_.\})$. For example, take $\{S^i_.\} = \{K(Z/p^{\oplus n},1);\ n>0\}$ and $\{T^j_.\} = K(Z/p,2)$, where $K(Z/p^{\oplus n},1) \to K(Z/p^{\oplus n-1},1)$ is projection onto the first $n-1$ factors. Then $\pi_1(\text{Hom.}(\{K(Z/p^{\oplus n},1)\},K(Z/p,2))) = \bigoplus_{n>1} Z/p$;

$\pi_1(\text{Hom.}(\underset{\leftarrow}{\text{holim}}\{K(Z/p^{\oplus n},1)\},K(Z/p,2))) = \text{Hom}(\Pi\, Z/p, Z/p)$.

Propositions 13.4, 13.6 and 13.8 are intended to be used in conjunction with the comparison theorems of Chapter 8. The following proposition is typical of various consequences of these propositions. We remind the reader that $\#(\ ) = \{\text{cosk}_n(\ );\ n>0\}$.

PROPOSITION 13.10. *Let* X. *be a pointed, locally noetherian simplicial scheme, let* $\ell \neq p$ *be distinct primes, let* F *be the algebraic closure of* F_p, *let* k *be an algebraically closed field containing* F, *and let* R *denote the Witt vectors of* F. *For any complex reductive Lie group* $G(C) = G$, *a choice of embedding of* R *into* C *determines a homotopy class of homotopy equivalences of connected components*

(13.10.1) $(\underset{\leftarrow}{\text{holim}}\ \text{Hom.}((X.)_{et}, \#\circ(Z/\ell)_\infty(BG_k)_{et}))_0$

$$\sim (\underset{\leftarrow}{\text{holim}}\ \text{Hom.}((X.)_{et}, \#\circ\{(Z/\ell)_n\}\circ\text{Sing.}(BG)))_0\ .$$

If $f: Z. \to X.$ *is a map of pointed, locally noetherian simplicial schemes such that* $f^*: H^*(X., Z/\ell) \to H^*(Z., Z/\ell)$ *is an isomorphism, then*

$$(13.10.2)\quad f^*: \operatorname{holim}_{\leftarrow} \operatorname{Hom.}((X.)_{et}, \#\circ(Z/\ell)_\infty(BG_k)_{et}) \to \operatorname{holim}_{\leftarrow} \operatorname{Hom.}((Z.)_{et}, \#\circ(Z/\ell)_\infty(BG_k)_{et})$$

is a homotopy equivalence. If $H^i(X., Z/\ell)$ *is finite for all* $i \geq 0$, *then there is a natural homotopy class of homotopy equivalences as in (13.10.1) (but of not necessarily connected simplicial sets)*

$$(13.10.3)\quad \operatorname{holim}_{\leftarrow} \operatorname{Hom.}((X.)_{et}, \#\circ(Z/\ell)_\infty(BG_k)_{et}) \sim \operatorname{holim}_{\leftarrow} \operatorname{Hom.}((X.)_{et}, \#\circ\{(Z/\ell)_n\}\circ \operatorname{Sing.}(BG))$$

and a natural isomorphism for each $n \geq 0$

$$(13.10.4)\quad \pi_n(\operatorname{holim}_{\leftarrow} \operatorname{Hom.}((X.)_{et}, \#\circ(Z/\ell)_\infty(BG_k)_{et}) \simeq \operatorname{Hom}_{\text{pro-}\mathcal{H}_*}(\Sigma^n(X.)_{et}, \#\circ(Z/\ell)_\infty(BG_k)_{et})$$

(where the homotopy groups are those based at the point map). If $X.$ *is of finite type over* $\operatorname{Spec} \mathbb{C}$, *then there is a natural homotopy class of homotopy equivalences*

$$(13.10.5)\quad \operatorname{holim}_{\leftarrow} \operatorname{Hom.}((X.)_{et}, \#\circ(Z/\ell)_\infty(BG_{\mathbb{C}})_{et}) \sim \operatorname{Hom.}(\operatorname{Sing.}|X.^{top}|, (Z/\ell)_\infty \circ \operatorname{Sing.}(BG)) .$$

Proof. By Propositions 6.10 and 8.8, the chain of strict maps relating $(BG_k)_{et}$ and Sing. (BG) determine a chain of strict maps relating $\#\circ(Z/\ell)_\infty\circ(BG_k)_{et}$ and $\#\circ\{(Z/\ell)_n\}\circ\operatorname{Sing.}(BG)$ which are isomorphisms in pro-$\mathcal{H}_*$. Propositions 6.9 and 13.4 imply that this chain induces the homotopy equivalence of (13.10.1). The homotopy equivalence of (13.10.2) is implied by Proposition 13.6, which proves that $\operatorname{Hom.}((X.)_{et}, T^j.) \to \operatorname{Hom.}((Z.)_{et}, T^j.)$ is a homotopy equivalence for each j, where

$\# \circ (\mathbf{Z}/\ell)_\infty (BG_F)_{et} = \{T^j_.\}$. The homotopy equivalence (13.10.3) follows from the chain of strict maps relating $\# \circ (\mathbf{Z}/\ell)_\infty \circ (BG_k)_{et}$ and $\# \circ \{(\mathbf{Z}/\ell)_n\} \circ \mathrm{Sing.}\,(BG)$ by employing the pro-finite property of the homotopy pro-groups proved in Corollary 13.9 together with Proposition 6.9. This pro-finite property enables us to apply Corollary 13.9 to obtain isomorphism (13.10.4).

To obtain the homotopy equivalence of (13.10.5), we employ (13.10.3) to obtain a natural isomorphism in $\mathcal{H}_*$ relating $\underleftarrow{\mathrm{holim}}\ \mathrm{Hom.}\,((X.)_{et}, \# \circ (\mathbf{Z}/\ell)_\infty (BG_{\mathbf{C}})_{et})$ to $\underleftarrow{\mathrm{holim}}\ \mathrm{Hom.}\,((X.)_{et}, \# \circ \{(\mathbf{Z}/\ell)_n\} \circ \mathrm{Sing.}\,(BG))$ which in turn is related by a natural isomorphism in $\mathcal{H}_*$ to $\underleftarrow{\mathrm{holim}}\ \mathrm{Hom.}\,(\mathrm{Sing.}\,|X^{top}_.|, \# \circ \{(\mathbf{Z}/\ell)_n\} \circ \mathrm{Sing.}\,(BG))$ by Theorem 8.4 and Corollary 13.7. The homotopy equivalence (13.10.5) is now obtained from the natural isomorphism of [12], XI. 3.3

$$\underleftarrow{\mathrm{holim}}\ \mathrm{Hom.}\,(S.\,, \{T^j_.\}) \simeq \mathrm{Hom.}\,(S.\,, \underleftarrow{\mathrm{holim}}\,\{T^j_.\})$$

applied to the case $S. = \mathrm{Sing.}\,|X^{top}_.|$ and $\{T^j_.\} = \# \circ \{(\mathbf{Z}/\ell)_n\} \circ \mathrm{Sing.}\,(BG)$ together with the natural equivalence

$$(\mathbf{Z}/\ell)_\infty \circ \mathrm{Sing.}\,(BG) \to \underleftarrow{\mathrm{holim}} \circ \# \circ \{(\mathbf{Z}/\ell)_n\} \circ \mathrm{Sing.}\,(BG)\,. \blacksquare$$

We briefly explain how Propositions 13.2 and 13.10 are applied to the algebraic K-theory of rings A of finite type over an algebraically closed field k (and more generally to schemes quasi-projective over k). One verifies that the algebraic function complex $\mathrm{Hom.}\,(\mathrm{Spec}\,A \amalg \mathrm{Spec}\,k, BGL_{n,k})$ of Proposition 13.2 (where $GL_{n,k}$ is the algebraic group over k associated to $GL_n(\mathbf{C})$) equals $BGL_n(A)$. Thus Proposition 13.2 yields a map

$$(13.11) \quad BGL_n(A) \to \underleftarrow{\mathrm{holim}}\ \mathrm{Hom.}\,((\mathrm{Spec}\,A)_{et} \amalg \mathrm{pt}, \# \circ (\mathbf{Z}/\ell)_\infty (BGL_{n,k})_{et})\,.$$

Such maps for $n \geq 0$ determine a map from the algebraic K-theory of A (which can be defined to be the homotopy groups of the group completion of $\amalg BGL_n(A)$) to homotopy groups of a space constructed from the right-hand side of (13.11) for $n \geq 0$. Using Proposition 13.10, one can show that these latter homotopy groups are closely related to topological K-groups. See [39] for further details.

REMARK 13.12. In order to apply Proposition 13.10 to (13.11), one must consider function complexes of the form

$$\mathrm{Hom}.(\{S_.^i\} \amalg \mathrm{pt}., \{T_.^j\}); \quad \{S_.^i\}, \{T_.^j\} \in \text{pro-(s. sets}_{*c}) .$$

One views this function complex as the *function complex of not necessarily pointed maps* from $\{S_.^i\}$ to $\{T_.^j\}$. One readily verifies that the hypotheses of Proposition 13.6 also imply that

$$f^* : \mathrm{Hom}.(\{S_.^i\} \amalg \mathrm{pt}., T.) \to \mathrm{Hom}.(\{R_.^k\} \amalg \mathrm{pt}., T.)$$

is a homotopy equivalence. Consequently, Corollary 13.7, Proposition 13.8, Corollary 13.9, and Proposition 13.10 remain valid when one replaces all function complexes by associated function complexes of not necessarily pointed maps. ■

14. RELATIVE COHOMOLOGY

In this chapter, we introduce the relative (etale) cohomology of a map of simplicial schemes. As shown in Corollary 14.5, this relative cohomology provides an algebraic description of the relative cohomology of a pair of etale topological types, generalizing the relationship between the (etale) cohomology of a simplicial scheme and the cohomology of the etale topological type. The relative cohomology groups themselves are of particular interest in the study of duality theorems as well as other contexts in which a global question can be usefully reduced to a local context.

In Proposition 14.1, we recall the usual formulation of relative cohomology for an open or closed immersion of schemes. Our definition of relative cohomology for a general map of simplicial schemes given in Proposition 14.3 utilizes the mapping cylinder construction to replace this general map by a closed immersion. Proposition 14.6 verifies that this construction gives the "correct" definition, namely that of local cohomology, in the case of an open immersion. We conclude this chapter by constructing various products in relative cohomology and homology.

We begin by recalling certain adjoint functors used in the literature to formulate relative cohomology for open or closed immersions of schemes.

Let $i: Y \to X$ be a closed immersion of schemes with Zariski open complement $j: U \to X$. We define

$$i^! : \mathrm{AbSh}(X) \to \mathrm{AbSh}(Y)$$

for any $F \in \mathrm{AbSh}(X)$ and any separated $W \to Y$ in $\mathrm{Et}(Y)$ by setting $i^!F(W \to Y)$ to be the subgroup of $F(V \to X)$ consisting of sections with supports on W whenever $V \to X$ in $E(X)$ satisfies $V \times_Y X = W$. We define

$$j_! : \mathrm{AbSh}(U) \to \mathrm{AbSh}(X)$$

by defining $j_!F$ to be the sheaf associated to the presheaf $j_!^pF$ sending $V \to X$ to $\bigoplus_{\mathrm{Hom}_X(V,U)} F(V \to U)$.

Then $i^!$ is an exact right adjoint to i_*, since i is a closed immersion; and $j_!$ is an exact left adjoint to j^*, since j is etale (cf. [5], 2.5.). The subsheaf $i_*i^!F$ of F (also denoted $\underline{\Gamma}_Y F$) is called the subsheaf of sections with *supports on* Y ; the subsheaf $j_*j^!F$ of F is called the subsheaf of sections with *supports off* Y, where $j^!$ is right adjoint to j_*.

The relevance of these definitions to relative cohomology is seen in the following proposition.

PROPOSITION 14.1. *Let* $i: Y \to X$ *be a closed immersion of schemes with Zariski open complement* $j: U \to X$ *and let* F *be an abelian sheaf on* $\mathrm{Et}(X)$. *Then there are canonical exact sequences*

(14.1.1) $$0 \to i_*i^!F \to F \to j_*j^*F$$

(14.1.2) $$0 \to j_!j^*F \to F \to i_*i^*F \to 0 .$$

Furthermore, there are long exact sequences

(14.1.3) $$\cdots \to H^k_Y(X,F) \to H^k(X,F) \to H^k(U,j^*F) \to \cdots$$

(14.1.4) $$\cdots \to H^k(X,j_!j^*F) \to H^k(X,F) \to H^k(Y,i^*F) \to \cdots$$

where $H^k_Y(X,F)$ *is the* k-th *right derived functor of the functor*

$$H^0_Y(\ ) = H^0(X,\ ) \circ i_* \circ i^!$$

sending a sheaf G *to the subgroup of* $G(X)$ *consisting of sections with support on* Y.

Proof. The maps of sheaves in (14.1.1) and (14.1.2) are adjunction morphisms, the injectivity of $i_*i^!F \to F$ and $j_!j^*F \to F$ is seen by inspection, and the exactness of (14.1.1) and (14.1.2) is proved by checking exactness at stalks. The long exact sequence (14.1.4) is immediate from (14.1.2) (because i_* is exact and sends injectives to injectives, $H^*(X,i_*i^*F)$ equals $H^*(Y,i^*F)$).

If $F \to I^\cdot$ is an injective resolution, then the exactness of (14.1.1) for each I^n implies the exactness of

(14.1.5) $$0 \to H^0(X,i_*i^!I^\cdot) \to H^0(X,I^\cdot) \to H^0(X,j_*j^*I^\cdot) \to 0$$

except possibly for the surjectivity of

$$H^0(X,I^\cdot) \to H^0(X,j_*j^*I^\cdot) = H^0(U,j^*I^\cdot) .$$

This surjectivity is immediate from the observation that every injective sheaf is flasque. Because $H^*(H^0(X,i_*i^!I^\cdot))$ equals $H^*_Y(X,F)$ (by definition) and $H^*(H^0(U,j^*I^\cdot))$ equals $H^*(U,j^*F)$ (because j^* is exact and sends injectives to injectives, thanks to the existence of its exact left adjoint $j_!$), we conclude that (14.1.5) implies (14.1.3). ∎

With (14.1.4) as a guide, we present the following definition.

DEFINITION 14.2. Let $i: Y. \to X.$ be a closed immersion of simplicial schemes, F an abelian sheaf on $Et(X.)$, G an abelian sheaf on $Et(Y.)$, and $\alpha : F \to i_*G$ a surjective map. Then we define $H^*(X.\ ,Y.\ ;F,G)$ (with α implicit) by

$$H^*(X.\ ,Y.\ ;F,G) = H^*(X.\ ,\ker(\alpha)) .$$

In particular, if $G = i^*F$ and α is the adjunction morphism, then we denote $H^*(X.\ ,Y.\ ;F,i^*F)$ by $H^*(X.\ ,Y.\ ;F)$, so that

$$H^*(X.\ ,Y.\ ;F) = H^*(X.\ ,j_!j^*F)$$

whenever $Y. \to X.$ admits a Zariski open complement $j: U. \to X.$ ∎

So defined, $H^*(X.\ ,Y.\ ;F,G)$ fits in a natural long exact sequence

(14.2.1) $$\cdots \to H^*(X.\ ,Y.\ ;F,G) \to H^*(X.\ ,F) \to H^*(Y.\ ,G) \to \cdots .$$

Moreover, if $0 \to F \to F' \to F'' \to 0$ is a short exact sequence in AbSh(X.), $0 \to i^*F \to i^*F' \to i^*F'' \to 0$ is also short exact so that

$$0 \to \mathrm{Ker}(a) \to \mathrm{Ker}(a') \to \mathrm{Ker}(a'') \to 0$$

is short exact (where a, a', a'' are the adjunction morphisms). Consequently, we obtain another natural long exact sequence

(14.2.2) $\cdots \to H^*(X.\,,Y.\,,F) \to H^*(X.\,,Y.\,;F') \to H^*(X.\,,Y.\,;F'') \to \cdots$.

If $f: Z. \to X.$ is a map of simplicial scheme, then the *mapping cylinder* of $f, M(f)$, is the simplicial scheme.

$$M(f) = Z. \otimes \Delta[1] \bigvee_{Z.\otimes\{1\}} X.$$

defined to be the natural quotient of $Z. \otimes \Delta[1] \amalg X.$ with

$$M(f)_n = Z_n \otimes (\Delta[1]_n - \{(1,\cdots,1)\}) \amalg X_n .$$

There is a canonical closed (and open) immersion $i: Z. \to M(f)$ identifying Z_n with $Z_n \otimes \{(0,\cdots,0)\}$, a canonical projection $p: M(f) \to X.$ with

$$p_n = f_n \amalg f_n \amalg \cdots \amalg f_n \amalg 1 : M(f)_n \to X_n ,$$

and a canonical section $s: X. \to M(f)$.

Using this mapping cylinder construction to replace an arbitrary map by a closed immersion, we present our general definition of relative cohomology (following P. Deligne's definition in [23], 6.3.2.1 for simplicial spaces). While defining relative cohomology in general, we must verify that our definition agrees with Definition 14.2 for closed immersions.

PROPOSITION 14.3. *Let* $f: Z. \to X.$ *be a map of simplicial schemes,* F *an abelian sheaf on* Et(X.), G *an abelian sheaf on* Et(Z.), *and* $a: F \to f_*G$ *a map of sheaves on* Et(X.). *Let* $M(a)$ *denote the abelian sheaf on* $M(f)$ *determined by* a *with* $M(a)_n$ *restricted to each copy of* Z_n *given by* G_n *and restricted to* X_n *given by* F_n, *so that* $M(a) = p^*F$

if $\alpha : F \to f_* f^* F$ *is the adjunction morphism. We define the relative cohomology groups* $H^*(X.\ ,Z.\ ; F,G)$ *by*

$$H^*(X.\ ,Z.\ ;F,G) = H^*(M(f),Z; M(\alpha)) = H^*(M(f),\ker(\beta))$$

where $\beta : M(\alpha) \to i_*G = i_*i^*M(\alpha)$ *is the adjunction morphism. If* f *is a closed immersion and* α *is surjective, then there is a natural isomorphism*

$$p^* : H^*(X.\ ,\ker(\alpha)) \to H^*(M(f),\ker(\beta))$$

so that this definition agrees with that of Definition 14.2. For $f : Z. \to X.$ *and* $\alpha : F \to f_*G$ *arbitrary, there is a natural long exact sequence*

(14.3.1) $$\cdots \to H^*(X.\ ,Z.\ ; F,G) \to H^*(X.\ ,F) \to H^*(Z.\ ,G) \to \cdots .$$

Moreover, if we let $H^*(X.\ ,Y.\ ; F)$ *be defined by*

$$H^*(X.\ ,Z.\ ; F) = H^*(X.\ ,Z.\ ; F, f^*F) ,$$

then a short exact sequence $0 \to F \to F' \to F'' \to 0$ in AbSh(X.) *determines the long exact sequence*

(14.3.2) $$\cdots \to H^*(X.\ ,Z.\ ; F) \to H^*(X.\ ,Z.\ ; F') \to H^*(X.\ ,Z.\ ; F'') \to \cdots .$$

Proof. We first verify that $p : M(f) \to X.$ and the natural map $F \to p_*M(\alpha)$ induce an isomorphism

(14.3.3) $$p^* : H^*(X.\ ,F) \xrightarrow{\sim} H^*(M(f),M(\alpha)) .$$

The section $s : X. \to M(f)$ determines a left inverse s^* for p^*. Moreover, there exists a homotopy $h : M(f) \otimes \Delta[1] \to M(f)$ relating 1 and $s \circ p$ and a natural map $M(\alpha) \to h_*pr^*M(\alpha)$ (where $pr : M(f) \otimes \Delta[1] \to M(f)$ is the projection map) determining $h^* : H^*(M(f),M(\alpha)) \to H^*(M(f) \otimes \Delta[1], pr^*M(\alpha))$ such that

$$i_1^* \circ h^* = 1 ,\quad i_2^* \circ h^* = p^* \circ s^* : H^*(M(f),M(\alpha)) \to H^*(M(f),M(\alpha))$$

where $i_1, i_2 : M(f) \to M(f) \otimes \Delta[1]$ are the canonical inclusions. Consequently, Lemma 14.4 implies (14.3.3).

The long exact sequence (14.3.1) follows immediately from (14.2.1) and (14.3.3). The long exact sequence (14.3.2) is obtained as was (14.2.2) above. Because $p: M(f) \to X.$ restricts to f on $Z.$, the natural map $F \to p_* M(\alpha)$ restricts to $\ker(\alpha) \to p_* \ker(\beta)$ and thus induces

$$p^* : H^*(X. ,\ker(\alpha)) \to H^*(M(f),\ker(\beta)) .$$

If f is a closed immersion and $\alpha: F \to f_* G$ is surjective, then the 5-Lemma together with (14.2.1) and (14.3.1) implies that this map is an isomorphism. ∎

LEMMA 14.4. *Let* $X.$ *be a simplicial scheme and* F *an abelian sheaf on* $Et(X.)$. *Then the projection map* $p: X. \otimes \Delta[1] \to X.$ *and the adjunction morphism* $F \to p_* p^* F$ *induce an isomorphism*

$$p^* : H^*(X. ,F) \xrightarrow{\simeq} H^*(X. \otimes \Delta[1],p^*F)$$

whose inverse is given by both $i_1 : X. \to X. \otimes \Delta[1], p^*F \to i_{1*}F$ *and* $i_2 : X. \otimes \Delta[1], p^*F \to i_{2*}F$.

Proof. Because $p \circ i_\varepsilon = 1$ for $\varepsilon = 0$ or 1, we obtain $p^*F \to i_{\varepsilon *}F$ as the adjoint of the identity map $1 : F \to F$. Observe that

$$U.. \otimes \Delta[1] \to X. \otimes \Delta[1] = p^*(U.. \to X.)$$

is a hypercovering of $X. \otimes \Delta[1]$ whenever $U.. \to X.$ is a hypercovering of $X.$. As argued in the proof of Theorem 3.8, we conclude the natural isomorphism

$$H^*(X. \otimes \Delta[1],p^*F) \xrightarrow{\simeq} \operatorname{colim} H^*(p^*F(\Delta U.. \otimes \Delta[1]))$$

where the colimit is indexed by $U..$ in $HR(X.)$. Consequently, the lemma follows from Theorem 3.8 and the fact that

$$p^* \circ i_\varepsilon^* : p^*F(\Delta U.. \otimes \Delta[1]) \to F(\Delta U..) \to p^*F(\Delta U.. \otimes \Delta[1])$$

is cochain homotopic to the identity. ∎

If $g: S. \to T.$ is a map of simplicial sets, then the mapping cylinder of $g, M(g)$, is defined by

$$M(g) = (S. \times \Delta[1]) \bigvee_{S. \times \{1\}} T. .$$

If L is a local coefficient system on $T.$, then $H^*(T. ,S. ;L)$ is defined by

$$H^*(T. ,S. ; L) = H^*(C^{\cdot}(M(g),S. ; p^*L))$$

where

$$C^{\cdot}(M(g),S. ; p^*L) = \mathrm{Ker}(p^*L(M(g)) \to g^*L(S.))$$

and where $p: M(g) \to T.$ is the canonical projection. If g is an inclusion, then $H^*(T. ,S. ; L)$ is naturally isomorphic to $H^*(C^{\cdot}(T. ,S. ; L))$. If

$$g: \{S^i_{\cdot} ; i \in I\} \to \{T^j_{\cdot} ; j \in J\}$$

is a map of pro-simplicial sets and if L is a local coefficient system on $\{T^j_{\cdot}\}$, then

$$H^*(\{T^j_{\cdot}\}, \{S^i_{\cdot}\}; L) = \mathrm{colim}\, H^*(T^j_{\cdot} ,S^i_{\cdot} ; L)$$

is well defined, where the colimit is indexed by all maps $S^i_{\cdot} \to T^j_{\cdot}$ representing $g_j: \{S^i_{\cdot}\} \to T^j_{\cdot}$ for some $j \in J$. Moreover, $H^*(\{T^j_{\cdot}\}, \{S^i_{\cdot}\}; L)$ fits in a natural long exact sequence

(14.5.1) $$\cdots \to H^*(\{T^j_{\cdot}\}, \{S^i_{\cdot}\}; L) \to H^*(\{T^j_{\cdot}\},L) \to H^*(\{S^i_{\cdot}\},g^*L) \to \cdots .$$

Since the construction in Proposition 14.3 is the scheme theoretic analogue of the above construction of relative cohomology of simplicial sets, the following corollary is an easy consequence of Proposition 14.3.

COROLLARY 14.5. *Let* $f: Z. \to X.$ *be a map of locally noetherian simplicial schemes and* L *a locally constant sheaf on* $\mathrm{Et}(X.)$. *There is a natural isomorphism*

$$H^*(X. ,Z. ; L) \xrightarrow{\sim} H^*((X.)_{et},(Z.)_{et}; L)$$

which, together with the isomorphisms of Proposition 5.9, fit in an isomorphism of long exact sequences arising from (14.3.2) and (14.5.1).

Proof. By definition,

$$H^*((X.)_{et},(Z.)_{et};L) = \operatorname{colim} H^*(\pi(\Delta U..),\pi(\Delta W..);p^*L)$$

where the colimit is indexed by $g: W.. \to U..$ in $HR(f)$. If L restricted to $U_{0,0}$ is constant, then

$$C^{\cdot}(M(\pi(\Delta g)),\pi(\Delta W..);p^*L) = \operatorname{Ker}(\beta)(\Delta M(g))$$

where $M(g)$ is a hypercovering of $M(f)$ and $\beta : p^*L \to i_*f^*L$ is the natural map. Consequently, the functor $HR(f) \to HR(M(f))$ sending $g: W.. \to U..$ to $M(g)$ determines a map of long exact sequences

$$\begin{array}{ccccccc} \cdots \to H^*((X.)_{et},(Z.)_{et};L) & \to & H^*(\{M(\pi(\Delta g))\},p^*L) & \to & H^*(\{\pi(\Delta W..)\},f^*L) & \to & \cdots \\ \downarrow & & \downarrow & & \downarrow & & \\ \cdots \to \operatorname{colim} H^*(\operatorname{Ker}(\beta)(\Delta Q..)) & \to & \operatorname{colim} H^*(p^*L(\Delta Q..)) & \to & \operatorname{colim} H^*(i_*f^*L(\Delta Q..)) & \to & \cdots \end{array} \qquad (14.5.2)$$

The left finality of $s: HR(f) \to HR(Z)$ implies that the top row of (14.5.2) can be identified with (14.5.1); the bottom row of (14.5.2) (whose colimits are indexed by $Q.. \in HR(M(f))$) can be identified with (14.3.1) thanks to (14.3.3). The asserted isomorphism now follows by applying the 5-Lemma to (14.5.2) in conjunction with Proposition 5.9. ■

The long exact sequences (14.1.3) and (14.3.1) suggest the following proposition, a version of which was first proved by D. Cox (cf. Propositions 1.6 and 1.8 of [21]). The cohomology groups $H^*_{Y.}(X. , \ )$ are often called *local cohomology* groups.

PROPOSITION 14.6. *Let* $j: U. \to X.$ *be a Zariski open inclusion of simplicial schemes which is the complement of some closed immersion* $i: Y. \to X.$. *For each* $t \geq 0$,

$$H^t(X. ,U. ;\ ) : \mathrm{AbSh}(X.) \to (\text{ab. grps})$$

is the t-th *right derived functor* $H^t_{Y.}(X. ,\ )$ *of the functor* $H^0_{Y.}(X. ,\ )$ *sending* F *to* $\ker(H^0(X. ,F) \to H^0(U. ,j^*F))$.

Proof. By (14.3.1), $H^0(X. ,U. ;\ ) = H^0_{Y.}(X. ,\ )$. Consequently, (14.3.2) implies that it suffices to prove that $H^t(X. ,U. ; I) = 0$ for $t > 0$ and I injective. Because $j : U. \to X.$ is the complement of some closed immersion, the functors $j_{n!} : \mathrm{AbSh}(U_n) \to \mathrm{AbSh}(X_n)$ determine a functor $j_! : \mathrm{AbSh}(U.) \to \mathrm{AbSh}(X.)$ left adjoint to j^*. The exactness of $j_!$ implies that j^* sends injectives to injectives. The injectivity of $j_! j^* Z \to Z$ in AbSh(X.) implies the surjectivity of $H^0(X. ,I^\cdot) \to H^0(U. ,j^*I^\cdot)$. Therefore, (14.3.1) implies that $H^t(X. ,U. ,I) = 0$ for $t > 0$ and I injective. ∎

It is useful to keep in mind that products are defined in cohomology. In particular, the cup product multiplication is an immediate consequence of the following *external product multiplication.*

PROPOSITION 14.7. *Let* $f : Y. \to X.$ *and* $g : W. \to Z.$ *be closed immersions of simplicial schemes and let* $\mu : p_1^*F \otimes p_2^*G \to H$ *be a map in* $\mathrm{AbSh}(X. \times Z.)$, *where* $p_1 = pr_1 : X. \times Z. \to X.$ *and* $p_2 = pr_2 : X. \times Z. \to Z.$. *Then* μ *determines a natural pairing*

$$\times : H^*(X. ,Y. ; F) \otimes H^*(Z. ,W. ; G) \to H^*(X. \times Z. ,(X. \times W.) \cup (Y. \times Z.); H) .$$

Proof. Because a section $s \otimes t$ with the support of s off Y. and the support of t off W. has support off $(X. \times W.) \cup (Y. \times Z.) = T.$, μ induces a map

$$p_1^* \ker(F \to f_* f^* F) \otimes p_2^* \ker(G \to g_* g^* G) \to \ker(H \to h_* h^* H)$$

where $h = (1 \times g) \cup (f \times 1) : T. \to X. \times Z.$. By replacing F, G, H by these kernels, we may assume Y. and W. are empty.

For any $0 \le p \le n$, let $\lambda_p = d_{p+1} \circ \cdots \circ d_n : X_n \to X_p$ and let $\rho_{n-p} = d_0 \circ \cdots \circ d_0 : Z_n \to Z_{n-p}$. We employ the Alexander-Whitney map

$$F_p(U_{p,p}) \otimes G_{n-p}(V_{n-p}, V_{n-p}) \to \lambda_p^* F_p(U_{n,n}) \otimes \rho_{n-p}^* G_{n-p}(V_{n,n})$$
$$\to F_n(U_{n,n}) \otimes G_n(V_{n,n})$$

together with the natural map determined by μ

$$F_n(U_{n,n}) \otimes G_n(V_{n,n}) \to (p_1^* F \otimes p_2^* G)_n(U_{n,n} \times V_{n,n})$$
$$\to H_n(U_{n,n} \times V_{n,n})$$

to obtain a pairing of chain complexes

$$F(\Delta U..) \otimes G(\Delta V..) \to H(\Delta(U.. \times V..))$$

for any $U.. \in HR(X.), V.. \in HR(Z.)$. Taking colimits and using the canonical maps

$$H^*(F(\Delta U..)) \otimes H^*(G(\Delta V..)) \to H^*(F(\Delta U..) \otimes G(\Delta V..)) ,$$

we obtain the required pairing

$$\times : H^*(X. ,F) \otimes H^*(Z. ,G) \to H^*(X. \times Z. ,H) . \blacksquare$$

We state the following cup product pairing for pro-objects of simplicial schemes in order that it be applicable to tubular neighborhoods as required in Chapter 17. As usual, the cohomology of a pro-object (of simplicial schemes or pairs of simplicial schemes) is the direct limit of the cohomology of the constituent objects.

COROLLARY 14.8. *Let* $\{f^i : Y^i_. \to X^i_. \; ; i \in I\}$ *be a pro-object of pairs of simplicial schemes. Then there is a natural cup product pairing for each* $p,q \geq 0$

$$\cup : H^p(\{X^i_.\}, F) \otimes H^q(\{X^i_.\}, \{Y^i_.\}; G) \to H^{p+q}(\{X^i_.\}, \{Y^i_.\}; F \otimes G)$$

for any pair of abelian sheaves F, G *defined on some* $Et(X^j_.)$.

Proof. Replacing $X^i_.$ by $M(f^i)$, we may assume each f^i is a closed immersion. For $\phi : i \to j$ in I, the pairing

(14.8.1) $$H^p(X^i_\cdot, F) \otimes H^q(X^i_\cdot, Y^i_\cdot; G) \to H^{p+q}(X^i_\cdot, Y^i_\cdot; F \otimes G)$$

is obtained from that of Proposition 14.7 by composing with

$$H^{p+q}(X^i_\cdot \times X^i_\cdot, X^i_\cdot \times Y^i_\cdot; p_1^* F \otimes p_2^* G) \to H^{p+q}(X^i_\cdot, Y^i_\cdot; F \otimes G)$$

induced by the diagonal $\Delta: (X^i_\cdot, Y^i_\cdot) \to (X^i_\cdot \times X^i_\cdot, X^i_\cdot \times Y^i_\cdot)$ (in an abuse of notation, we let F and G denote $\phi^* F$ and $\phi^* G$ on $Et(X^i_\cdot)$). The asserted pairing is the colimit with respect to $i \in I$ of (14.8.1). ∎

To define relative homology, we simply dualize the definition of relative cohomology given in Proposition 14.3.

DEFINITION 14.9. Let $f: Z. \to X.$ be a map of simplicial schemes and let F be an abelian sheaf on Et(X.). We define the *relative homology* of (X. ,Z.) with coefficients F^V by

$$H_*(X., Z.; F^V) = H_*(M(f), \ker(\beta)^V)$$

where $\beta: p^* F \to i_* f^* F$ is the canonical surjection in AbSh(M(f)) (see Proposition 14.3). ∎

As in Corollary 14.5, we conclude that

(14.9.1) $$H_*(X., Z.; M^V) \simeq H_*((X.)_{et}, (Z.)_{et}; M^0)$$

whenever $f: Z. \to X.$ is a map of noetherian simplicial schemes and M is locally constant and constructible (see also the proof of Proposition 7.6).

In order to exhibit cup products in the generality required for Chapter 17, we require the following lemma (implying Mayer-Vietoris for $X. = U. \cup V.$).

LEMMA 14.10. *Let X. be a locally noetherian simplicial scheme and let $U. \to X.$, $V. \to X.$ be Zariski open immersions whose union is X. and whose intersection we denote by W. . Then there is a natural map in* pro-(s. sets)

$$\Pi = (U.)_{et} \underset{(W.)_{et}}{\vee} (V.)_{et} \to (X.)_{et}$$

which induces an isomorphism in cohomology with coefficients determined by any locally constant abelian sheaf M *on* Et(X.) .

Proof. We re-index (cf. [8], A.3.3) to obtain a pro-object of commutative squares of simplicial sets

$$\begin{array}{ccc} (W.)_{et} & \longrightarrow & (U.)_{et} \\ \big\downarrow & & \big\downarrow \\ (V.)_{et} & \longrightarrow & (X.)_{et} \end{array} \;\simeq\; \left\{ \begin{array}{ccc} S^j_\cdot & \longrightarrow & R^j_\cdot \\ \big\downarrow & & \big\downarrow \\ Q^j_\cdot & \longrightarrow & T^j_\cdot \end{array} \right\} = \{D^j\} .$$

We define Π to be $\{R^j_\cdot \vee_{S^j_\cdot} Q^j_\cdot\}$, where $R^j_\cdot \vee_{S^j_\cdot} Q^j_\cdot$ is the diagonal of the bi-simplicial set

$$sk_1(R^j_\cdot \amalg Q^j_\cdot \leftleftarrows R^j_\cdot \amalg S^j_\cdot \amalg Q^j_\cdot \cdots)$$

(where $sk_1(\ )$ refers to the "horizontal" simplicial structure).

Let N.. denote the Čech nerve of the Zariski open covering $U. \amalg V. \to X.$, so that Proposition 8.1 asserts that the natural map $(\Delta N..)_{et} \to (X.)_{et}$ is a weak equivalence. This map clearly factors through $\Pi \to (X.)_{et}$ via a map $(\Delta N..)_{et} \to \Pi$ which is easily checked to induce an isomorphism in cohomology. ■

Lemma 4.10 is useful because $R. \vee_{S.} Q.$ is naturally equivalent to $R'_\cdot \cup Q'_\cdot$ with $R'_\cdot \cup Q'_\cdot = S.$, where $R'_\cdot = R. \amalg_{S.} S. \times \Delta[1] \sim R.$ and $Q'_\cdot = S. \times \Delta[1] \amalg_{S.} Q. \sim Q.$.

PROPOSITION 14.11. *Let* $\{U^i_\cdot \to X^i_\cdot\,;\, V^i_\cdot \to X^i_\cdot\,;\, i \in I\}$ *be a pro-object of pairs of Zariski open inclusions of locally noetherian simplicial schemes. For each* $n \geq p \geq 0$, *there exists a natural cap product pairing of abelian groups*

$$\cap : H^p(\{(X^i_\cdot)_{et}\},\{(U^i_\cdot)_{et}\};M) \otimes \lim_{\leftarrow} H_n(\{(X^i_\cdot)_{et}\},\{(U^i_\cdot \cup V^i_\cdot)_{et}\};N^0)$$
$$\to \lim_{\leftarrow} H_{n-p}(\{(X^i_\cdot)_{et}\},\{(V^i_\cdot)_{et}\};(M \otimes N)^0) .$$

for any pair of locally constant abelian sheaves M, N *on some* $Et(X^i_.)$.

Proof. By applying Lemma 14.10 to each $U^i_. \cup V^i_.$ and by applying the mapping cylinder construction to the composition

$$(U^i_.)_{et} \underset{(U^i_. \cap V^i_.)_{et}}{\vee} (V^i_.)_{et} \to (U^i_. \cup V^i_.)_{et} \to (X^i_.)_{et} ,$$

we are reduced to exhibiting a cap product pairing

$$\begin{aligned} &\operatorname{colim} H^p(T^j_., R^j_. ; M) \otimes \underset{\leftarrow}{\lim}\, H_n(T^j_., R^j_. \cup S^j_. ; N) \\ (14.11.1) \qquad & \qquad \to \underset{\leftarrow}{\lim}\, H_{n-p}(T^j_., S^j_. ; M \otimes N) \end{aligned}$$

for a pro-object of pairs of inclusions of simplicial sets $\{R^j_. \to T^j_., S^j_. \to T^j_.\}$. For given inclusions $R. \to T.$, $S. \to T.$, the cap product

(14.11.2) $\quad H^p(T., R. ; M) \otimes H_n(T., R. \cup S. ; N) \to H_{n-p}(T., S. ; M \otimes N)$

sends

$$\alpha \in \prod_{y \in T_p} M(y) , \qquad \beta \in \bigoplus_{x \in T_n} N(x)$$

to

$$\alpha \cap \beta = \sum_{t \in T_{n-p}} \ \sum_{\rho_{n-p}(x)=t} \alpha(\lambda_p(x)) \otimes \beta(x) \in \bigoplus_{t \in T_{n-p}} M(t) \otimes N(t)$$

where $\lambda_p = d_0 \circ \cdots \circ d_0 : T_n \to T_p$ and $\rho_{n-p} = d_{n-p+1} \circ \cdots \circ d_n : T_n \to T_{n-p}$. If $\phi : T'_. \to T.$ restricts to $R'_. \to R.$ and $S'_. \to S.$, then

$$\phi_*(\phi^* \alpha \cap \beta') = \alpha \cap \phi_*(\beta')$$

for any $\alpha \in H^p(T., R. ; M)$ and $\beta' \in H_n(T'_., R'_. \cup S'_. ; N)$. This equality suffices to guarantee that the product (14.11.2) for each pair of inclusions $R^j_. \to T^j_.$, $S^j_. \to T^j_.$ determines (14.11.1). ∎

15. TUBULAR NEIGHBORHOODS

Although examples indicate that there may be no (global) etale neighborhoods of a closed subscheme $Y \subset X$, there are always sufficiently many (local) etale neighborhoods about a given geometric point of Y. This observation is exploited in D. Cox's thesis in which he constructs etale tubular neighborhoods of closed immersions (cf. [18]). In this chapter, we present a somewhat simplified version of Cox's construction generalized to simplicial schemes and refined to topological types.

The fundamental property of the tubular neighborhood $T_{X./Y.}$ of a closed immersion $Y. \to X.$ is that $T_{X./Y.}$ has the same homotopy type as $Y.$ (as shown in Theorem 15.4). As we see in Proposition 15.5, excision properties are conveniently stated in terms of tubular neighborhoods. In Proposition 15.6, we describe $X.$ homotopy theoretically as a union of $T_{X./Y.}$ and the complement of $Y.$ with intersection the deleted tubular neighborhood $T_{X./Y.} - Y.$. We conclude this chapter with a theorem of Cox's asserting that the deleted tubular neighborhoods of a closed immersion of smooth varieties and of the associated normal bundle minus the zero section determine weakly equivalent sphere fibrations.

DEFINITION 15.1. Let $f: Y. \to X.$ be a pointed, closed immersion of locally noetherian simplicial schemes. Define

$$\psi_{Y.}(\ ): \mathrm{Et}(X.) \to \mathrm{Et}(X.)$$

by setting $\psi_{Y.}(U \to X_n)$ equal to the union of those connected components U^α of U with $U^\alpha \times_{X_n} Y_n \neq \emptyset$. We define the *tubular neighborhood* of $Y.$ in $X.$, $T_{X./Y.}$, to be the following pro-object of pointed simplicial schemes

$$T_{X./Y.} = \{\psi_{Y.}(\Delta U..); U.. \in HRR(X.)\} .$$

We define the category of abelian sheaves on $T_{X./Y.}$, $AbSh(T_{X./Y.})$, by

$$AbSh(T_{X./Y.}) = \operatorname{colim} AbSh(\psi_{Y.}(\Delta U..))$$

where the colimit is indexed by $U.. \in HRR(X.)$. Finally, we define the "weak tubular neighborhood" of $Y.$ in $X.$, $t_{X./Y.}$, to be the following pro-object in the homotopy category of pointed simplicial schemes

$$t_{X./Y.} = \{\psi(\Delta U..); U.., u \in HR(X., x)\} . \blacksquare$$

We proceed to prove that both $T_{X./Y.}$ and $t_{X./Y.}$ have cohomology isomorphic to that of $Y.$. In the context of schemes, a somewhat restricted version of this result for $t_{X./Y.}$ was originally proved by D. Cox ([18], Theorem 1.3).

PROPOSITION 15.2. *Let* $f: Y. \to X.$ *be a (pointed) closed immersion of locally noetherian simplicial schemes. Let* $F \in AbSh(T_{X./Y.})$ *be such that* f^*F *descends to* $f^*F \in AbSh(Y.)$. *Then there are natural isomorphisms*

$$\begin{aligned} H^*(T_{X./Y.}, F) &\equiv \operatorname*{colim}_{HRR(X.)} H^*(\psi_{Y.}(\Delta U..), F) \\ &\simeq \operatorname*{colim}_{HR(X.)} H^*(\psi_{Y.}(\Delta U..), F) \equiv H^*(t_{X./Y.}, F) \\ &\simeq \operatorname*{colim}_{HR(X.)} H^*(F(\psi_{Y.}(\Delta U..)) \\ &\simeq H^*(Y., f^*F) . \end{aligned}$$

Proof. As in the proof of Corollary 4.6, the first isomorphism follows from Propositions 3.4 and 4.5. The proof of Theorem 3.8 applies with only minor changes to prove the second isomorphism. To prove the third isomorphism, we consider some $V.. \in HR(X.)$ and let $g: T. \to W.$ denote the closed immersion $\Delta V.. \times_{X.} Y. \to \psi_{Y.}(\Delta V..)$. If $F \in AbSh(W.)$, then the natural map $F \circ \psi_{Y.} \to g_* g^* F = f_* f^* F$ of abelian presheaves on $Et(W.)$ induces an isomorphism of associated sheaves, as can be verified by inspection at each stalk. Therefore, Corollary 3.10 implies the isomorphism

$$\operatorname{colim} H^*(F(\psi_{Y.}(\Delta U..))) \xrightarrow{\sim} \operatorname{colim} H^*(f_* f^* F(\Delta U..)) .$$

Consequently, the third isomorphism follows from Theorem 3.8 and the isomorphism

$$H^*(X. , f_* f^* F) \simeq H^*(Y. , f^* F) . \blacksquare$$

Because $H^*(X. , Y. ; F)$ is defined to be $H^*(X. , \ker(F \to f_* f^* F))$ and because the sheaf associated to the presheaf $\ker(F \to F \circ \psi_{Y.})$ is isomorphic to $\ker(F \to f_* f^* F)$ by the above argument, Corollary 3.10 immediately implies the following corollary.

COROLLARY 15.3. *Let* $f : X. \to Y.$ *be a closed immersion of locally noetherian simplicial schemes and let* F *be an abelian sheaf on* $Et(X.)$. *Then there is a natural isomorphism*

$$H^*(X. , Y. ; F) \simeq \operatorname*{colim}_{HR(X.)} H^*(\ker(F(\Delta U..) \to F \circ \psi_{Y.}(\Delta U..))) .$$

In particular, if M *is a locally constant abelian sheaf on* $Et(X.)$, *then*

$$H^*(X. , Y. ; M) \simeq \operatorname*{colim}_{HR(X.)} H^*(\pi(\Delta U..), \pi(\psi_{Y.}(\Delta U..)); M) . \blacksquare$$

We next compare the homotopy type of the tubular neighborhood $T_{X./Y.}$ with that of $Y.$. Once again, the basic case of the weak equivalence $(Y.)_{ht} \to (t_{X./Y.})_{ht}$ is due to D. Cox ([18], Theorem 2.2).

THEOREM 15.4. *Let* $f : Y. \to X.$ *be a pointed closed immersion of locally noetherian simplicial schemes. We define*

$$(T_{X./Y.})_{et} = \{\pi(\psi_{Y.}(\Delta U..)); U.. \in HRR(X.)\} \in \text{pro-(s. sets}_*)$$

$$(t_{X./Y.})_{ht} = \{\pi(\psi_{Y.}(\Delta U..)); U.. \in HR(X.)\} \in \text{pro-}\mathcal{H}_* .$$

As in Corollary 6.3, there is a natural commutative diagram in pro-$\mathcal{H}_*$

(15.4.1)
$$\begin{array}{ccccc} (Y.)_{ht} & \longrightarrow & (t_{X./Y.})_{ht} & \longrightarrow & (X.)_{ht} \\ \downarrow & & \downarrow & & \downarrow \\ (Y.)_{et} & \longrightarrow & (T_{X./Y.})_{et} & \longrightarrow & (X.)_{et} \end{array}$$

whose vertical arrows are weak equivalences, whose lower horizontal arrows are strict maps in pro-(s. sets$_*$), *and whose horizontal compositions are equal to* f_{et} *and* f_{ht}. *Furthermore,* $(Y.)_{ht} \to (t_{X./Y.})_{ht}$ *and* $(Y.)_{et} \to (T_{X./Y.})_{et}$ *are also weak equivalences.*

Proof. The maps in (15.4.1) are the natural ones induced by the functors

$$\begin{array}{ccccc} HR(Y.) & \longleftarrow & HR(X.) & = & HR(X.) \\ \uparrow & & \uparrow & & \uparrow \\ HRR(Y.) & \longleftarrow & HRR(X.) & = & HRR(X.) \end{array} .$$

The maps $(Y.)_{ht} \to (Y.)_{et}$ and $(X.)_{ht} \to (X.)_{et}$ are weak equivalences by Corollary 6.3. (The connectivity hypothesis can be dropped by arguing as in the proof of Proposition 10.2.) By Proposition 15.2, $(t_{X./Y.})_{ht} \to (T_{X./Y.})_{et}$ induces a bijection of connected components (cf. Proposition 5.2). Using Proposition 5.6 as in the proof of Corollary 5.7, we conclude that $(t_{X./Y.})_{ht} \to (T_{X./Y.})_{et}$ induces an isomorphism of fundamental pro-groups of corresponding connected components. (As in the proof of Proposition 10.2, we use the left finality of $HR(X. , \{y_\alpha\}) \to HR(X.)$, where y_α is a chosen geometric point of the connected component $Y.^\alpha$ of $Y.$.) As argued in the proof of Corollary 5.8, abelian local coefficient systems on $(t_{X./Y.})_{ht}$ and $(T_{X./Y.})_{et}$ naturally correspond to locally constant abelian sheaves in $AbSh(T_{X./Y.})$. Consequently, using

Proposition 5.9 and Theorem 6.2, we conclude that Proposition 15.2 implies that $(t_{X./Y.})_{ht} \to (T_{X./Y.})_{et}$ is isomorphic to a disjoint union of weak equivalences in pro-$\mathcal{H}_{*c}$.

To complete the proof of the theorem, it suffices to prove that $(Y.)_{ht} \to (t_{X./Y.})_{ht}$ is a weak equivalence. As argued above, this map is isomorphic to a disjoint union of maps in pro-$\mathcal{H}_{*c}$. We claim that HR(X.) contains a left final, full subcategory C of HR(X.) such that $U.. \in C$ implies that

$$\pi(\Delta U.. \underset{X.}{\times} Y.) = \pi(\psi_{Y.}(\Delta U..)) .$$

Then the image of C under $f^* : HR(X) \to HR(Y)$ satisfies the hypotheses of Corollary 5.7, so that $(Y.)_{ht} \to (t_{X./Y.})_{ht}$ induces an isomorphism of fundamental pro-groups of corresponding components. Consequently, Proposition 15.2 in conjunction with Proposition 5.9 and Theorem 6.2 implies that $(Y.)_{ht} \to (t_{X./Y.})_{ht}$ is isomorphic to a disjoint union of weak equivalences in pro-$\mathcal{H}_{*c}$.

To prove the claim, it suffices to prove for any $U.. \in HR(X.)$ that there exists a map $U..' \to U..$ in HR(X.) with $\pi(\Delta U..' \underset{X.}{\times} Y.) = \pi(\psi_{Y.}(\Delta U..))$. Let $\phi : Et(X.) \to Et(X.)$ be defined by sending $U \to X_n$ to

$$\phi(U) = \coprod_{\alpha} (U - \coprod_{\beta \neq \alpha} V^{\beta})$$

where $U \underset{X_n}{\times} Y_n = \coprod_{\alpha} V^{\alpha}$ with each V^{α} connected, non-empty. So defined, there is a natural transformation $\eta : \phi \to 1$ such that $\phi(U) \underset{X_n}{\times} Y_n \xrightarrow{\sim} U \underset{X_n}{\times} Y_n$, $\pi(U \underset{X_n}{\times} Y_n) = \pi(\psi_{Y.}(\phi(U)))$, and $\phi(U) \to U$ is etale, surjective. We define $\eta : U..' \to U..$ inductively as follows. Namely, $U'_{s,0} \to U_{s,0} = \phi(U_{s,0}) \to U_{s,0}$;

$$U'_{s,t} \to U_{s,t} = \phi((\cosk_{t-1} U'_{s.})_t \underset{(\cosk_{t-1} U_{s.})_t}{\times} U_{s,t}) \to U_{s,t} .$$

In verifying that $U..' \in HR(X.)$, the reader should employ the fact that ϕ is idempotent to obtain the degeneracy map $\sigma_j : U'_{s,t-1} \to U'_{s,t}$ from

$$(\sigma_j, \sigma_j \circ \eta) : U'_{s,t-1} \to (\operatorname{cosk}_{t-1} U'_{s.})_t \underset{(\operatorname{cosk}_{t-1} U_{s.})_t}{\times} U_{s,t} \, .$$

The other structure maps of $U..'$ are immediate, as is the fact that $U..' \to X.$ is a hypercovering. Finally, the equality $\pi(\Delta U..' \underset{X.}{\times} Y.) = \pi(\psi_{Y.}(U..'))$ is immediate from the corresponding equality $\pi(U \underset{X_n}{\times} Y_n) = \pi(\psi_{Y.}(\phi(U)))$ for ϕ. ■

In view of Corollary 15.3 and Theorem 5.4, we are justified in defining the *etale topological type of the closed immersion* $Y. \to X.$ by

$$(15.4.2) \qquad (X.\,,Y.)_{et} = \{\pi(\Delta U..), \pi(\Delta \psi_{Y.}(U..)); U.. \in HRR(X.)\} \, .$$

Using tubular neighborhoods, we can prove the following useful excision isomorphisms. The first, (15.5.1), asserts that $H^*(X.\,,Y.\,;F)$ can be computed on any $Z.$ containing the complement of $Y.$; the second, (15.5.2), asserts that $H^*(X.\,,X. - Z.\,;F)$ can be computed by restricting to the tubular neighborhood of $Y.$ in $X.$.

PROPOSITION 15.5. *Let* $f : Y. \to X.$ *be a closed immersion of locally noetherian simplicial schemes and let* $j : Z. \to X.$ *be a Zariski open immersion such that* $X_n - Y_n \subset Z_n$ *for each* $n \geq 0$. *For any abelian sheaf* F *on* $Et(X.)$, *there exist excision isomorphisms*

$$(15.5.1) \quad H^*(X.\,,Y.\,;F) \simeq H^*(T_{X./X.} \underset{X.}{\times} Z.\,, T_{X./Y.} \underset{X.}{\times} Z.\,; j^*F)$$

$$(15.5.2) \qquad H^*(X.\,,Z.\,;F) \simeq H^*(T_{X./Y.}, T_{X./Y.} \underset{X.}{\times} Z.\,;F) \, .$$

Proof. We first verify for any $U.. \in HR(X.)$ that j induces an isomorphism

$$(15.5.3) \quad H^*(\Delta U..\,, \psi_{Y.}(\Delta U..); F) \xrightarrow{\sim} H^*(\Delta U.. \underset{X.}{\times} Z.\,, \psi_{Y.}(\Delta U..) \underset{X.}{\times} Z.\,; j^*F) \, .$$

By Proposition 2.4, to prove (15.5.3) it suffices to verify that j_n induces an isomorphism

$$(15.5.3)_n \quad H^*(U_{n,n},\psi_{Y_n}(U_{n,n});F_n) \xrightarrow{\sim} H^*(U_{n,n} \underset{X_n}{\times} Z_n,\psi_{Y_n}(U_{n,n}) \underset{X_n}{\times} Z_n;j_n^*F_n)$$

for each $n \geq 0$. Because j_n^* is exact with an exact left adjoint $j_{n!}$ (cf. Proposition 14.1), both sides vanish on injectives and send short exact sequences of coefficient sheaves to long exact sequences of cohomology groups. Thus, it suffices to prove $(15.5.3)_n$ for $* = 0$. By (14.2.1), this is merely the statement

$$\ker(F_n(U_{n,n}) \to F_n(\psi_{Y_n}(U_{n,n}))) = \ker(j_n^*F_n(U_{n,n} \underset{X_n}{\times} Z_n) \to j_n^*F_n(\psi_{Y_n}(U_{n,n}) \underset{X_n}{\times} Z_n))$$

which is verified by inspection.

The natural map $(\Delta U_{..},\Delta U_{..} \underset{X.}{\times} Y.) \to (\Delta U_{..},\psi_{Y.}(\Delta U_{..}))$ induces the composition

$$\begin{aligned} H^*(\Delta U_{..},\psi_{Y.}(\Delta U_{..}); F) &\longrightarrow H^*(\Delta U_{..},\Delta U_{..} \underset{X.}{\times} Y. ; F) \\ &\xleftarrow{\sim} H^*(X.,Y. ; F) . \end{aligned}$$

By Proposition 15.2 together with (14.2.1), we conclude that these maps induce an isomorphism

$$\operatorname{colim} H^*(\Delta U_{..},\psi_{Y.}(\Delta U_{..}),F) \xrightarrow{\sim} H^*(X.,Y. ; F) .$$

Consequently, the colimit with respect to $U_{..} \in HRR(X.)$ of (15.5.3) yields (15.5.1).

In a similar fashion, we verify for any $U_{..} \in HR(X.)$ that $\psi_{Y.}(\Delta U_{..}) \to \Delta U_{..}$ induces an isomorphism (by Propositions 2.6 and 14.6)

$$(15.5.4) \quad H^*(\Delta U_{..},\Delta U_{..} \underset{X.}{\times} Z. ;F) \xrightarrow{\sim} H^*(\psi_{Y.}(\Delta U_{..}),\psi_{Y.}(\Delta U_{..}) \underset{X.}{\times} Z. ;F) .$$

Then the colimit with respect to $U_{..} \in HRR(X.)$ of (15.5.4) yields (15.5.2). ∎

In the following proposition, we verify that the etale topological type of $X.$ is homotopic to the homotopy theoretic union of the etale topological types of $T_{X./Y.}$, and of any $Z.$ containing the complement of $Y.$ ($X. - Y.$ may not be a simplicial scheme). In more geometric language, $X.$ is obtained from $Z.$ by sewing in $T_{X./Y.}$ along $T_{X./Y.} \underset{X.}{\times} Z.$.

The *homotopy theoretic push-out* of a diagram of pointed simplicial sets

$$R.\,,r \leftarrow S.\,,s \rightarrow T.\,,t$$

is the following pointed simplicial set (cf. Lemma 14.10)

$$R. \underset{S.}{\vee} T. = R. \underset{S.}{\cup} (S. \times \Delta[1]) \underset{S.}{\cup} T./\{r\} \cup (\{s\} \times \Delta[1]) \cup \{t\}.$$

The homology and cohomology of $R. \underset{S.}{\vee} T.$ are described by Mayer-Vietoris sequences, whereas $\pi_1(R. \underset{S.}{\vee} T.)$ is given by the Van Kampen theorem.

PROPOSITION 15.6. *Let* $f: Y. \rightarrow X.$ *be a closed immersion of connected, locally noetherian simplicial schemes and let* $j: Z. \rightarrow X.$ *be a Zariski open immersion of connected simplicial schemes such that* $X_n - Y_n \subset Z_n$ *for each* $n \geq 0$. *Choose a geometric point* x *of* Z_0 *some specialization of which is a geometric point of* Y_0. *Then the natural map in* pro-(s. sets$_{*c}$)

$$(15.6.1) \qquad (T_{X./Y.})_{et} \underset{(T_{X./Y.} \underset{X.}{\times} Z.)_{et}}{\vee} (Z.)_{et} \rightarrow (X.)_{et}$$

is a weak equivalence in pro-$\mathcal{H}_*$, *where*

$$(T_{X./Y.} \underset{X.}{\times} Z.)_{et} = \{(\psi_{Y.}(\Delta U..) \underset{X.}{\times} Z.)_{et};\ U.. \in HRR(X.)\}$$

Proof. For any $U.. \in HR(X.\,,x)$, we have the equality

$$(15.6.2) \qquad \psi_{Y.}(\Delta U..) \underset{(\psi_{Y.}(\Delta U..) \underset{X.}{\times} Z.)}{\cup} (\Delta U.. \underset{X.}{\times} Z.) = \Delta U.. \ ;$$

in other words, $\Delta U..$ is the union of the Zariski open subsimplicial schemes $\psi_{Y.}(\Delta U..)$, $\Delta U.. \underset{X.}{\times} Z.$ with intersection $\psi_{Y.}(\Delta U..) \underset{X.}{\times} Z.$.

We first observe that (15.6.2) induces a natural map in pro-(s. sets$_{*c}$)

$$(15.6.3)\quad (\psi_{Y.}(\Delta U..))_{et} \underset{(\psi_{Y.}(\Delta U..) \underset{X.}{\times} Z.)_{et}}{\vee} (\Delta U.. \underset{X.}{\times} Z.)_{et} \to (\Delta U..)_{et} .$$

Because an isomorphism class of principal G-fibrations over $\Delta U..$ is equivalent to a pair of isomorphism classes of principal G-fibrations over $\psi_{Y.}(\Delta U..)$ and $\Delta U.. \underset{X.}{\times} Z.$ whose restrictions to $\psi_{Y.}(\Delta U..) \underset{X.}{\times} Z.$ are equal, the Van Kampen theorem and Proposition 5.6 imply that (15.6.3) induces an isomorphism of fundamental groups.

Observe that there is a natural bijection

$$\Delta R.. \underset{\Delta S..}{\vee} \Delta T.. \to \Delta(R.. \underset{S..}{\vee} T..)$$

for any pair of maps $R.. \leftarrow S.. \to T..$ of bisimplicial sets. Consequently, the left finality of $HRR(V.) \to HRR(V_n)$ for any $V.$ and any n implies that in order to prove that (15.6.3) induces an isomorphism in cohomology with coefficients in M (corresponding to a locally constant abelian sheaf M on $Et(\Delta U..)$) it suffices to consider for each $n \geq 0$

$$(15.6.3)_n \qquad (U'_n)_{et} \underset{(U'_n \underset{X_n}{\times} Z_n)_{et}}{\vee} (U_{n,n} \underset{X_n}{\times} Z_n)_{et} \to (U_{n,n})_{et}$$

where $U'_n = \psi_{Y_n}(U_{n,n})$. The fact that $(15.6.3)_n$ induces an isomorphism in cohomology with coefficients M_n is immediate from the observation that

$$(U_{n,n} \underset{X_n}{\times} Z_n)_{et} = (U'_n \underset{X_n}{\times} Z_n)_{et} \amalg (V_n)_{et}, \quad (U_{n,n})_{et} = (U'_n)_{et} \amalg (V_n)_{et}$$

where $U_{n,n} = U'_n \amalg V_n$. Consequently, (15.6.3) is a weak equivalence.

To prove that the inverse system (indexed by $U.. \in HRR(X.)$) of weak equivalences (15.6.3) determines the weak equivalence (15.6.1), it suffices to prove that the natural map

$$(15.6.4)\quad \{(\psi_{Y.}(\Delta U..))_{et}; U.. \in HRR(X.)\} \to \{\pi(\psi_{Y.}(\Delta U..)); U.. \in HRR(X.)\} = (T_{X./Y.})_{et}$$

is a weak equivalence. This follows directly from Propositions 5.6 and 15.2. ■

We conclude this chapter with a sketch of the following result of D. Cox ([18], Theorems 3.2 and 5.1). Geometrically, this theorem asserts that the *deleted tubular neighborhood* of a closed immersion of smooth algebraic varieties determines the same "*normal spherical fibration*" as that determined by the normal bundle of the immersion (see Corollary 10.9).

THEOREM 15.7 (D. Cox). *Let* $f: Y \to X$ *be a closed immersion of codimension* c *of connected, smooth, quasi-projective varieties over an algebraically closed field* k. *Then the homotopy fibre of the map*

$$(Z/\ell)_\infty(T_{X/Y} - Y)_{et} \to (Z/\ell)_\infty(T_{X/Y})_{et}$$

is weakly equivalent to $\{(Z/\ell)_n(S^{2c-1}); n > 0\}$, *where*

$$T_{X/Y} - Y = T_{X/Y} \underset{X}{\times} (X - Y) .$$

Moreover, there exists a commutative diagram in pro-(s. sets$_{*c}$)

$$(15.7.1)\quad \begin{array}{ccccccc}
(Z/\ell)_\infty(T_{X/Y}-Y)_{et} & \to & (Z/\ell)_\infty(T_{X'/Y'}-Y')_{et} & \leftarrow & (Z/\ell)_\infty(T_{N/Y}-Y)_{et} & \to & (Z/\ell)_\infty(N-o(Y))_{et} \\
\downarrow & & \downarrow & & \downarrow & & \downarrow \\
(Z/\ell)_\infty(T_{X/Y})_{et} & \to & (Z/\ell)_\infty(T_{X'/Y'})_{et} & \leftarrow & (Z/\ell)_\infty(T_{N/Y})_{et} & \to & (Z/\ell)_\infty N_{et}
\end{array}$$

whose horizontal arrows are weak equivalences in pro-$\mathcal{H}_*$, *where* $o: Y \to N$ *is the zero section of the normal bundle*

$$N = \mathrm{Sym}_{O_Y}(\mathcal{J}_Y/\mathcal{J}_Y^2) \to Y \quad \textit{of} \quad f: Y \to X .$$

Proof. To prove that the homotopy fibres of the vertical maps of (15.7.1) are completed spheres, Cox employs the Spivak converse of the Thom isomorphism theorem ([15], 1.4.3) in the context of a map in pro-$\mathcal{H}_*$ and cohomology with $\mathbf{Z}/\ell$ coefficients. The case in which $c > 1$ is relatively straightforward; for $c = 1$, much more work must be done.

For (15.7.1), we consider the closed immersion $F : Y' = Y \times \mathbf{A}^1 \to X'$ (of codimension c of connected, smooth, quasi-projective varieties), the "deformation of the zero section of the normal bundle" constructed by Baum-Fulton-MacPherson for $k = \mathbf{C}$ ([10], I.5.1). There is a map $\pi : X' \to \mathbf{A}^1_k$ such that $F \circ \pi = pr_2 : Y \times \mathbf{A}^1 \to \mathbf{A}^1$,

$$Y'_0 = (F \circ \pi)^{-1}(0) \to \pi^{-1}(0) = X'_0$$

is isomorphic to $o : Y \to N$, and

$$Y'_t = (F \circ \pi)^{-1}(t) \to \pi^{-1}(t) = X'_t$$

is isomorphic to $f : Y \to X$ for any closed point $t \neq 0$.

The $\mathbf{Z}/\ell$-acyclicity of $\mathbf{A}^1_k$ and the fibres of $p : N \to Y$ (isomorphic to $\mathbf{A}^c_k$) imply that $Y' \leftarrow Y \to N$ induces weak equivalences in pro-$\mathcal{H}_*$ $(\mathbf{Z}/\ell)_\infty(Y')_{et} \leftarrow (\mathbf{Z}/\ell)_\infty(Y)_{et} \to (\mathbf{Z}/\ell)_\infty N_{et}$. Therefore, Theorem 15.4 implies that the lower row of (15.7.1) consists of weak equivalences. Using the maps between Serre spectral sequences associated to the vertical arrows of (15.7.1), we conclude that the upper row of (15.7.1) consists of weak equivalences provided that the induced map on homotopy fibres induces an isomorphism in $\mathbf{Z}/\ell$ cohomology. The cohomology of a fibre of a vertical arrow of (15.7.1) can be naturally identified with the stalk of the appropriate local cohomology sheaf (e.g., for the left vertical arrow, the cohomology of the fibre is the stalk of $R^{2c} f_* f^! \mathbf{Z}/\ell$). Naturality of these local cohomology sheaves implies the required isomorphisms of the cohomology of the fibres. ∎

16. GENERALIZED COHOMOLOGY

We present the definition and basic properties of the generalized (etale) cohomology groups of a closed immersion of simplicial schemes. Of most importance is the existence and convergence of an Atiyah-Hirzebruch type spectral sequence (16.5.1). As seen in Proposition 16.9, this spectral sequence can be reinterpreted in terms of a "local-to-global" spectral sequence. In the special case of complex K-theory with finite coefficients, this material has been recently applied to the study of algebraic K-theory; this is briefly described in Example 16.3.

The reader should recall (e.g., [3]) that a *generalized cohomology theory* is a contravariant functor

$$h^{\cdot}(\ ): \mathcal{H}^2 \to \text{gr. Ab}$$

from the homotopy category of pairs of simplicial sets to the category of graded abelian groups satisfying all the Eilenberg-Steenrod axioms except the dimension axiom. Such a theory is represented by an *Ω-spectrum* $\underline{\underline{E}} = \{E_k; \rho_k : \Sigma E_k \to E_{k+1}; k \in \mathbf{Z}\}$, where each E_k is a pointed Kan complex and the adjoint of each ρ_k is a pointed homotopy equivalence. Namely, for all pairs of integers k and m,

$$h^k(T.\ ,S.) = \mathrm{Hom}_{\mathcal{H}_0}(T./S.\ ,E_k) \simeq \mathrm{Hom}_{\mathcal{H}_0}(\Sigma^m(T./S.),E_{k+m}) .$$

We define $h^k(T.)$ by

$$h^k(T.) = h^k(T.\amalg \text{pt.}\ ,\text{pt}) .$$

DEFINITION 16.1. Let $h^{\cdot}(\ )$ be a generalized cohomology theory and let $\{T^i_\cdot\ ,S^i_\cdot\ ,i \in I\} \in \text{pro-}\mathcal{H}^2$. For any $k \in \mathbf{Z}$, we define

$$h^k(\{T^i_\cdot\ S^i_\cdot\}) = \mathrm{Hom}_{\text{pro-}\mathcal{H}_0}(\{T^i_\cdot/S^i_\cdot\}, \#(E_k))$$
$$\simeq \mathrm{Hom}_{\text{pro-}\mathcal{H}_0}(\{\Sigma^m(T^i_\cdot/S^i_\cdot)\}, \#(E_{k+m}))$$

where $\underline{\underline{E}} = \{E_m; m \in \mathbf{Z}\}$ represents $h^\cdot(\ )$ and where $\#(E_k) = \{\mathrm{cosk}_n E_k; n > 0\}$ (cf. Definition 6.1). If $Y_\cdot \to X_\cdot$ is a closed immersion of locally noetherian simplicial schemes, we define $h^k((X_\cdot, Y_\cdot)_{et})$ for any $k \in \mathbf{Z}$ (using (15.4.2)) by

$$h^k((X_\cdot, Y_\cdot)_{et}) = h^k(\{\pi(\Delta U_{\cdot\cdot}), \pi(\Delta\psi_{Y_\cdot}(U_{\cdot\cdot})); U_{\cdot\cdot} \in HRR(X_\cdot)\}) . \blacksquare$$

The consideration of maps into $\#(E_k)$ rather than E_k is dictated to us by the convergence of (16.4.1). The "ghost-classes" which are thereby eliminated appear extraneous in examples so far considered.

The following proposition may help to elucidate Definition 16.1.

PROPOSITION 16.2. *Let* $h^\cdot(\ )$ *be a generalized cohomology theory and let* $\{T^i_\cdot, S^i_\cdot; i \in I\} \in \text{pro-}\mathcal{H}^2$.

a.) *If* $\{T^i_\cdot, S^i_\cdot\} \simeq (T_\cdot, S_\cdot)$ *in* $\text{pro-}\mathcal{H}^2$ *and if either* $T_\cdot \supset S_\cdot$ *is finite dimensional or* $h^k(S^0, pt.) = h^k(pt.)$ *is finite for each* k, *then*

$$h^\cdot(\{T^i_\cdot, S^i_\cdot\}) \simeq h^\cdot(T_\cdot, S_\cdot) .$$

b.) *If* $\{L^j_\cdot, K^j_\cdot; j \in J\} \in \text{pro-(s. sets)}^2$ *with* $\{T^i_\cdot, S^i_\cdot\} \simeq \{L^j_\cdot, K^j_\cdot\}$ *in* $\text{pro-}\mathcal{H}^2$, *then*

$$h^\cdot(\{T^i_\cdot, S^i_\cdot\}) \simeq \lim_{\overleftarrow{n}} \operatorname*{colim}_{J} h^\cdot(sk_n L^j_\cdot, sk_n K^j_\cdot) .$$

c.) *If* $Y_\cdot \to X_\cdot$ *is a closed immersion of locally noetherian simplicial schemes, then*

$$h^\cdot((X_\cdot, Y_\cdot)_{et}) \simeq \lim_{\overleftarrow{n}} \operatorname*{colim}_{HRR(X_\cdot)} h^\cdot(sk_n\pi(\Delta U_{\cdot\cdot}), sk_n\pi(\Delta\psi_{Y_\cdot}(U_{\cdot\cdot}))) .$$

Proof. To prove a.), we utilize the isomorphism

$$h^k(\{T^i_\cdot, S^i_\cdot\}) \simeq \mathrm{Hom}_{\text{pro-}\mathcal{H}_0}(T_\cdot/S_\cdot, \#(E_k))$$

determined by the isomorphism $\{T^i_., S^i_.\} \simeq (T_., S_.)$. If $T_. \supset S_.$ is finite dimensional (i.e., $T_. = sk_n T_.$ for some n), then

$$\mathrm{Hom}_{\text{pro-}\mathcal{H}_0}(T_./S_., \#(E_k)) = \mathrm{Hom}_{\mathcal{H}_0}(T_./S_., E_k).$$

If $h^m(S^0, pt.) = \pi_{k+m}(E_k)$ is finite for all $m \geq -k$, then $\mathrm{Hom}_{\mathcal{H}_0}(\ \ , E_k)$ is a "compact representable functor" (cf. [69]); thus,

$$\mathrm{Hom}_{\mathcal{H}_0}(T_./S_., E_k) \simeq \varprojlim \mathrm{Hom}_{\mathcal{H}_0}(sk_n(T_./S_.), E_k) = \mathrm{Hom}_{\text{pro-}\mathcal{H}_0}(T_./S_., \#(E_k)).$$

Part b.) readily follows from the adjointness of $sk_n(\ )$ and $cosk_n(\ )$ together with the following observations: i.) if $sk_n S_. \rightrightarrows T_.$ are homotopic, then their adjoints $S_. \rightrightarrows cosk_n T_.$ are homotopic; and ii.) if $S_. \rightrightarrows cosk_n T_.$ are homotopic, then the restrictions of their adjoints $sk_{n-1} S_. \rightrightarrows T_.$ are homotopic. Finally, c.) is a special case of b.). ∎

EXAMPLE 16.3. Let $m > 1$ be an integer and let $K^{\cdot}(\ \ ; \mathbf{Z}/m)$ denote the generalized cohomology theory of mod-m, periodic, complex K-theory. Then $K^{\cdot}(\ \ ; \mathbf{Z}/m)$ is represented by the Ω-spectrum $\underline{\underline{BU}}^{C(m)}$, where $\underline{\underline{BU}}$ is the periodic spectrum of complex K-theory and $C(m)$ is the mod-m Moore space describable as the mapping cone of multiplication by m on the circle.

In Theorem 3.4 of [38], a natural transformation

$$\overline{\rho}_0 : K_0^{alg}(X,Y) \to K^0((X,Y)_{et}; \mathbf{Z}/m)$$

is defined for pairs (X,Y), where X is a quasi-projective algebraic variety over an algebraically closed field k containing $1/m$ and Y is a closed subvariety. In the special case $k = \mathbf{C}$, $\overline{\rho}_0$ is induced by the "forgetful functor" sending an algebraic vector bundle to its underlying topological vector bundle; in general, $\overline{\rho}_0$ factors the etale cohomological chern classes on $K_0^{alg}(X,Y)$. In Theorem 1.3 of [39], $\overline{\rho}_0$ is extended to higher degrees: natural maps

$$\bar{\rho}_i : K_i^{alg}(X,Y;\mathbf{Z}/m) \to K_i^{et}(X,Y;\mathbf{Z}/m) \simeq K^{-i}((X,Y)_{et},\mathbf{Z}/m)$$

are defined for all $i \geq 0$ and X quasi-projective over k (using 13.11).

These natural transformations have provided a useful means of studying *algebraic K-groups* of algebraic varieties over k. In fact, ambitious conjectures exist concerning the possibility that $\bar{\rho}_i$ might be an isomorphism under various hypotheses. For more general schemes, W. Dwyer and the author have recently developed in [27] a "twisted generalized cohomology theory" which appears to offer the correct generalization of $K^{\cdot}((\ )_{et},\mathbf{Z}/m)$. ■

The *Atiyah-Hirzebruch spectral sequence* is the standard tool for relating generalized cohomology to ordinary cohomology. When the *coefficients* of $h^{\cdot}(\ )$ (i.e., the groups $h^k(S^0,pt.) = h^k(pt.), k \in \mathbf{Z}$) are non-zero for arbitrarily large negative integers, then this spectral sequence need not converge.

PROPOSITION 16.4. *Let* $h^{\cdot}(\ )$ *be a generalized cohomology theory and let* $\{T^i_{\cdot}, S^i_{\cdot}; i \in I\} \in \text{pro-}\mathcal{H}^2$. *Then there is a natural half-plane spectral sequence (which is not necessarily convergent)*

$$\text{(16.4.1)} \qquad E_2^{p,q} = H^p(\{T^i_{\cdot}, S^i_{\cdot}\}; h^q) \Rightarrow h^{p+q}(\{T^i_{\cdot}, S^i_{\cdot}\}),$$

where $h^q = h^q(S^0,pt.)$. *This spectral sequence is strongly convergent if and only if*

$$\lim_{\substack{\leftarrow \\ r>p}}{}^1 E_r^{p,q} = 0, \quad all \ p \geq 0, \quad all \ q.$$

In particular, (16.4.1) is strongly convergent if $E_2^{p,q}$ *is finite for all* $p \geq 0$, *all* q; *or if for each* n, *there are only finitely many pairs* (p,q) *with* $p+q = n$ *and* $E_2^{p,q} \neq 0$ *(e.g.,* $E_2^{p,q} = 0$ *for* $p > N$, *or* $E_2^{p,q} = 0$ *for* $q < M$ *).*

Proof. We consider the following exact couple for each $i \in I$

$$(16.4.2)_i \qquad \begin{array}{ccc} \bigoplus_{p,q} h^{p+q}(sk_p T^i_\cdot, sk_p S^i_\cdot) & \longrightarrow & \bigoplus_{p,q} h^{p+q}(sk_{p-1} T^i_\cdot, sk_{p-1} S^i_\cdot) \\ & \nwarrow \qquad \swarrow & \\ & \bigoplus_{p,q} h^{p+q}(sk_p T^i_\cdot, sk_{p-1} T^i_\cdot \cup sk_p S^i_\cdot) . & \end{array}$$

Because the first derived exact couple of $(16.4.2)_i$, $(16.4.2)_i'$, is functorial on I (i.e., maps $(T^i_\cdot, S^i_\cdot) \rightrightarrows (T^j_\cdot, S^j_\cdot)$ equal in $\mathcal{H}^2$ induce the same map $(16.4.2)_i' \to (16.4.2)_j'$) and because $\operatorname{colim}_I(\ )$ is exact, we conclude that $\operatorname{colim}_I (16.4.2)_i'$ is an exact couple. To identify the spectral sequence associated to $\operatorname{colim}_I (16.4.2)_i'$ with (16.4.1), it suffices to observe that $d_1 : E_1^{p,q}(i) \to E_1^{p+1,q}(i)$ associated to $(16.4.2)_i$ is identified with the coboundary map of $C^\cdot(T^i_\cdot, S^i_\cdot; h^q)$ under the excision isomorphism

$$h^{p+q}(sk_p T^i_\cdot, sk_{p-1} T^i_\cdot \cup sk_p S^i_\cdot) \approx C^p(T^i_\cdot, S^i_\cdot; h^q) .$$

By definition, (16.4.1) *strongly converges* to $h^\cdot(\{X^i_\cdot, Y^i_\cdot\})$ if for all $p \geq 0$ and all q $\bigcap_{p<r} E_r^{p,q} = E_\infty^{p,q}$ is isomorphic to the p-th associated graded piece of $h^{p+q}(\{T^i_\cdot, S^i_\cdot\})$ filtered by the kernels of the maps

$$h^{p+q}(\{T^i_\cdot, S^i_\cdot\}) \to \operatorname{colim}_I \operatorname{im}\{h^{p+q}(sk_{p+1} T^i_\cdot, sk_{p+1} S^i_\cdot) \to h^{p+q}(sk_p T^i_\cdot, sk_p S^i_\cdot)\}$$

and if $h^{p+q}(\{T^i_\cdot, S^i_\cdot\})$ maps surjectively onto the inverse limit of its associated quotient filtration. Thanks to our use of $\#(E_k)$ in Definition 16.1, the convergence criterion of the proposition is a special case of Proposition A.2 of [38]. ∎

As an immediate corollary of Proposition 16.4 and Corollary 15.3, we conclude the following.

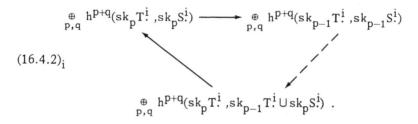

COROLLARY 16.5. *Let* $Y. \to X.$ *be a closed immersion of locally noetherian simplicial schemes and let* $h^{\cdot}(\ )$ *be a generalized cohomology theory. Then there is a natural spectral sequence*

$$(16.5.1) \qquad E_2^{p,q} = H^p(X.\ ,Y.\ ; h^q) \Longrightarrow h^{p+q}((X.\ ,Y.)_{et})$$

where $H^p(X.\ ,Y.\ ; h^q)$ *is the relative etale cohomology of* $Y. \to X.$ *with coefficients in the constant abelian sheaf* $h^q = h^q(S^0, pt.)$. *Moreover, the convergence discussion for (16.4.1) applies to (16.5.1).* ∎

The following corollary is a sample application of (16.5.1).

COROLLARY 16.6. *Let* $f: Y. \to X.$ *be a closed immersion of pointed, connected simplicial schemes of finite type over* $\mathbb{C}$, *and let* $h^{\cdot}(\ )$ *be a generalized cohomology theory with finite coefficients. Then there is a natural isomorphism*

$$h^{\cdot}((X.\ ,Y.)_{et}) \approx \lim_{\substack{\leftarrow \\ n}} h^{\cdot}(\mathrm{Sing.}(|sk_n X.^{top}|), \mathrm{Sing.}(|sk_n Y.^{top}|))\ .$$

Proof. The functor $HRR(X.) \to HRR(f)$ (cf. Definition 10.1) sending $U..$ to $f^*(U..) \to U..$ determines a map

$$\{\pi(\Delta U..), \pi(\Delta V..); V.. \to U.. \in HRR(f)\} \to \{\pi(\Delta U..), \pi(\Delta\psi_{Y.}(U..)); U.. \in HRR(X.)\}$$

which induces an isomorphism in cohomology (with locally constant sheaves on $Et(X.)$ as coefficients) by Proposition 15.2. By Theorem 8.4, the maps

$$\{\pi(\Delta U..), \pi(\Delta V..); V.. \to U.. \in HRR(f)\} \leftarrow \{\Delta \circ \mathrm{Sing.}(\Delta U..^{top}), \Delta \circ \mathrm{Sing.}(\Delta V..^{top}); V.. \to U.. \in HRR(f)\}$$
$$\to (\Delta \circ \mathrm{Sing.}(X.^{top}), \Delta \circ \mathrm{Sing.}(Y.^{top}))$$

induce isomorphisms in cohomology with h^q as coefficients, any $q \in \mathbf{Z}$. Moreover, each cohomology group $H^p(X.\ ,Y.\ ; h^q)$ is finite by Theorem 7.7, so that the above maps in pro-(s. sets)^2 determine maps of strongly convergent spectral sequences of type (16.4.1). We conclude that these maps induce isomorphisms on abutments. Finally, $\varprojlim_n h^{\cdot}(\mathrm{Sing.}\,|sk_n X^{top}_{.}|,$ $\mathrm{Sing.}\,|sk_n Y^{top}_{.}|)$ is the abutment of the spectral sequence for

$$(\Delta \circ \mathrm{Sing.}\,(X^{top}_{.}), \Delta \circ \mathrm{Sing.}\,(Y^{top}_{.}))\ ,$$

because the fact that each X_n is finite dimensional implies the existence of maps $S^n_{.} = \mathrm{Sing.}\,|sk_n X^{top}_{.}| \to sk_{n+f(n)} \circ \Delta \circ \mathrm{Sing.}\,(X^{top}_{.}) = T^{n+f(n)}_{.}$ such that the composites $S^n_{.} \to S^{n+f(n)}_{.}$, $T^n_{.} \to T^{n+f(n)}_{.}$ are equivalent to the natural inclusions. ∎

LEMMA 16.7. *Let* $Y. \to X.$ *and* $Z. \to Y.$ *be closed immersions of locally noetherian simplicial schemes, and let* $h^{\cdot}(\ )$ *be a generalized cohomology theory. Assume either that*

$$H^p(X.\,,Y.\,;h^q),\ H^p(X.\,,Z.\,;h^q),\ H^p(Y.\,,Z.\,;h^q)$$

are finite for all $p \geq 0$ *and all* q, *or that*

$$H^p(X.\,,Y.\,;h^q) = H^p(X.\,,Z.\,;h^q) = H^p(Y.\,,Z.\,;h^q) = 0$$

for all $q < M$ *(for some* M*) and* $p \geq 0$, *or that*

$$H^p(X.\,,Y.\,;h^q) = H^p(X.\,,Z.\,;h^q) = H^p(Y.\,,Z.\,;h^q) = 0$$

for all $p > N$ *(for some* N*) and all* q. *Then there is a natural long exact sequence*

$$(16.7.1)\qquad \cdots \to h^m((X.\,,Y.)_{et}) \to h^m((X.\,,Z.)_{et}) \to h^m((Y.\,,Z.)_{et}) \to \cdots .$$

Proof. We obtain (16.7.1) as the inverse limit of the following sequence of exact sequences

$$(16.7.2)_n\qquad \begin{aligned} \cdots &\to \mathrm{colim}\ h^m(sk_n\pi(\Delta U..), sk_n\pi(\Delta\psi_{Y.}(U..))) = A^m_n \\ &\to \mathrm{colim}\ h^m(sk_n\pi(\Delta U..), sk_n\pi(\Delta\psi_{Z.}(U..))) = B^m_n \\ &\to \mathrm{colim}\ h^m(sk_n\pi(\Delta\psi_{Y.}(U..)), sk_n\pi(\Delta\psi_{Z.}(U..))) = C^m_n \to \cdots \end{aligned}$$

where the colimits are indexed by $U.. \in HRR(X.)$. To conclude the isomorphism $h^m((Y.\,,Z.)_{et}) \to \varprojlim C_n^m$, we apply (16.4.1) to the natural cohomology equivalences (by Proposition 15.2)

$$\{\pi(\Delta V..),\pi(\Delta\psi_{Z.}(V..));V..\in HRR(Y.)\}\to\{\pi(\Delta\psi_{Y.}(U..)),\pi(\Delta\psi_{Z.}(U..));U..\in HRR(X.)\}\,.$$

Under the first set of hypotheses, exactness of $(16.7.1) = \varprojlim (16.7.2)_n$ follows from the finiteness of $(A_n^m)' = \mathrm{im}(A_{n+1}^m \to A_n^m)$, $(B_n^m)'$, and $(C_n^m)'$ given by $\mathrm{colim}\,(16.4.2)_i'$ for $\{\pi(\Delta U..),\, \pi(\Delta\psi_{Y.}(U..))\}$. Otherwise, $\mathrm{colim}\,(16.4.2)_i'$ implies that $\varprojlim_n A_n^m$ maps isomorphically onto $(A_n^m)'$ for $n >> m$; similarly, $\varprojlim_n B_n^m$ and $\varprojlim C_n^m$ map isomorphically onto $(B_n^m)'$ and $(C_n^m)'$ respectively for $m >> n$. These facts easily imply the exactness of $\varprojlim_n (16.7.2)_n$. ∎

Our next proposition, analagous to Proposition 2.4, relates the generalized cohomology of a simplicial scheme to that of each of its levels.

PROPOSITION 16.8. *Let* $f: Y. \to X.$ *be a closed immersion of simplicial schemes and let* $h^{\cdot}(\;)$ *be a generalized cohomology theory satisfying the conditions that* $H^m(X_p,Y_p;h^q)$ *is finite for all* $m \geq 0$, $p \geq 0$, q, *and is* 0 *for all* $m >> p$, *all* q. *Then there is a natural, strongly convergent spectral sequence*

$$(16.8.1) \qquad E_1^{p,q} = h^q((X_p,Y_p)_{et}) \Rightarrow h^{p+q}((X.\,,Y.)_{et})\,.$$

Proof. Define $f': Y.' \to X.'$ canonically mapping to $f: Y. \to X.$ by setting

$$X_n' = \coprod_{s\in S(n)} X_s\,, \quad Y_n' = \coprod_{s\in S(n)} Y_s$$

where $S(n)$ denotes the set of surjective, monotone increasing maps $s: (0,1,\cdots,n) \to (0,1,\cdots,k)$ (with $k \leq n$ variable) and where $X_s = X_k$ and $Y_s = Y_k$. The map of cochain complexes

$$\{p \mapsto H^m(X_p, Y_p; h^q)\} \to \{p \mapsto H^m(X'_p, Y'_p; h^q)\}$$

is a cochain equivalence for each $m \geq 0$ and each q (cf. [57], Corollary 22.3), so that $H^*(X.\,, Y.\,; h^q) \to H^*(X'.\,, Y'.\,; h^q)$ is an isomorphism by Proposition 2.4. As argued in the proof of Corollary 16.6, (16.4.1) implies the isomorphism $h^*((X.\,, Y.)_{et}) \xrightarrow{\sim} h^*((X'.\,, Y'.)_{et})$.

Lemma 16.7 applies to provide the following exact couple

$$(16.8.2)\qquad \begin{array}{ccc} \bigoplus_{p,q} h^{p+q}((sk_p X'.,sk_p Y'.)_{et}) & \longrightarrow & \bigoplus_{p,q} h^{p+q}((sk_{p-1} X'.,sk_{p-1} Y'.)_{et}) \\ & \nwarrow \qquad \swarrow & \\ & \bigoplus_{p,q} h^{p+q}((sk_p X'.,sk_{p-1} X'. \cup sk_p Y'.)_{et}) & \end{array}$$

whose associated spectral sequence

$$(16.8.3)\quad E_1^{p,q} = h^{p+q}((sk_p X'.,sk_{p-1} X'. \cup sk_p Y'.)_{et}) \Longrightarrow \lim_{\substack{\leftarrow \\ p}} h^{p+q}((sk_p X'.,sk_p Y'.)_{et})$$

converges by the finiteness of each $E_1^{p,q}$ (which is verified using (16.5.1) and our hypotheses on $H^m(X_p, Y_p, h^q)$). Using obstruction theory as in Proposition 13.6, we conclude the existence of maps $(sk_p X'.)_{et} \to sk_{p+n(p)}(X'.)_{et}$ such that the composites

$$(sk_p X'.)_{et} \to (sk_{p+n(p)} X'.)_{et}\,,\quad sk_p(X'.)_{et} \to sk_{p+n(p)}(X'.)_{et}$$

are equivalent to the natural inclusions. These maps imply that the abutment $\lim\limits_{\substack{\leftarrow \\ p}} h^{p+q}((sk_p X'.\,, sk_p Y'.)_{et})$ of (16.8.3) is isomorphic to $h^{p+q}((X'.\,, Y'.)_{et}) \simeq h^{p+q}((X.\,, Y.)_{et})$.

Thus, to obtain (16.8.1) from (16.8.3), it suffices to prove the isomorphism

$$(16.8.4)\qquad h^q((X_p, Y_p)_{et}) \simeq h^{p+q}((sk_p X'.\,, sk_{p-1} X'. \cup sk_p Y'.)_{et})\,.$$

Proposition 4.7 in conjunction with Corollary 16.5 implies the isomorphism

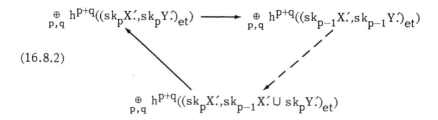

$$h^q((X_p,Y_p)_{et}) \simeq h^{p+q}((X_p \otimes \Delta[p], X_p \otimes sk_{p-1}\Delta[p] \cup Y_p \otimes \Delta[p])_{et}) .$$

Consequently, it suffices to prove that the natural map

$$(X_p \otimes \Delta[p], X_p \otimes sk_{p-1}\Delta[p] \cup Y_p \otimes \Delta[p]) \to (sk_p X.', sk_{p-1}X.' \cup sk_p Y.')$$

induces an isomorphism in cohomology (with h^q as coefficients) in order to prove (16.8.4). This is proved by applying excision (Proposition 15.5) to each of the natural maps

$$(X_p \otimes \Delta[p]_n, X_p \otimes (sk_{p-1}\Delta[p])_n \cup Y_p \otimes \Delta[p]_n) \to ((sk_p X.')_n, (sk_{p-1}X.' \cup sk_p Y.')_n) ,$$

because the isomorphism $X_p \simeq X_p' - (sk_{p-1}X.')_p$ implies the isomorphism $X_p \otimes (\Delta[p]_n - (sk_{p-1}\Delta[p])_n) \simeq (sk_p X.')_n - (sk_{p-1}X.')_n$. ∎

We conclude this chapter by proving that the Atiyah-Hirzebruch spectral sequence can be obtained as a *"local-to-global"* or *"descent" spectral sequence.* If one considered only Zariski hypercoverings of X. , (16.9.1) would become

$$E_2^{p,q} = H^p_{Zar}(X. ,Y. ; \underset{\sim}{h}^q_{Zar}) \Longrightarrow h^{p+q}((X. ,Y.)_{et})$$

where $H^*_{Zar}(\ )$ denotes Zariski cohomology and $\underset{\sim}{h}^q_{Zar}$ denotes the sheaf on the Zariski site associated to the presheaf $U \mapsto h^q(U_{et})$. The algebraic K-theory analogue of the above spectral sequence (for the Zariski site) is due to K. S. Brown and S. M. Gersten [16] (cf. [70] for a sophisticated algebraic K-theory analogue for the etale site).

PROPOSITION 16.9. *Let* $Y. \to X.$ *be a closed immersion of simplicial schemes of finite type over an algebraically closed field* k *and let* $h^{\bullet}(\ )$ *be a generalized cohomology theory with finite coefficients of orders invertible in* k. *Consider the following spectral sequences of type (16.8.1) indexed by noetherian hypercoverings* $U.. \in nHR(X.)$

$$E_1^{p,q}(\Delta U..) = h^q((U_{p,p}, \psi_{Y_p}(U_{p,p}))_{et}) \Longrightarrow h^{p+q}((\Delta U.., \Delta\psi_{Y.}(U..))_{et}) .$$

Then the colimit of these spectral sequences is well defined from the E_2*-level onward. This colimit spectral sequence can be written*

$$(16.9.1) \qquad E_2^{p,q} = H^p(X.\,,Y.\,;\underset{\sim}{h}^q) \Longrightarrow h^{p+q}((X.\,,Y.)_{et})$$

where $\underset{\sim}{h}^q$ *is the sheaf on* $Et(X.)$ *associated to the presheaf sending* $U \to X_n$ *in* $Et(X.)$ *to* $h^q(U_{et})$. *Moreover,* $\underset{\sim}{h}^q$ *is the constant sheaf* $h^q = h^q(S^0)$ *for all* q *and (16.9.1) is isomorphic to (16.5.1).*

Proof. The pair $(\Delta U..\,,\Delta\psi_{Y.}(U..))$ satisfies the hypotheses of Proposition 16.8 for any $U.. \in nHR(X.)$ by Theorem 7.7. The functorality of $U.. \mapsto \{E_r^{p,q}(\Delta U..);\, r \geq 2\}$ on $nHR(X.)$ follows from the observation that the first derived exact couple of (16.8.2) is functorial on (simplicial) homotopy classes of maps of pairs of simplicial schemes (exactly as functorality for $(16.4.2)_i'$). By Corollary 15.3 generalized to presheaves (exactly as Corollary 3.10 generalizes Theorem 3.8), we obtain isomorphisms

$$\operatorname{colim} E_2^{p,q}(\Delta U..) \simeq H^p(X.\,,Y.\,;\underset{\sim}{h}^q)\,.$$

Applying (16.4.1) to the map

$$(16.9.2) \quad \{(\Delta U..,\Delta\psi_{Y.}(U..))_{et};U.. \in nHR(X.)\} \to \{\pi(\Delta U..),\pi(\Delta\psi_{Y.}(U..));U.. \in nHR(X.)\}$$

and using Proposition 15.2 we conclude the isomorphisms

$$\operatorname{colim} h^{p+q}((\Delta U..\,,\Delta\psi_{Y.}(U..))_{et}) \simeq h^{p+q}((X.\,,Y.)_{et})\,.$$

Thus, (16.9.1) is obtained as the colimit of $U.. \mapsto \{E_r^{p,q}(\Delta U..);\, r \geq 2\}$.

To prove that $\underset{\sim}{h}^q$ is constant, it suffices to prove that the natural map

$$\operatorname{colim} h^q(V_{et}) \to h^q((\operatorname{Spec}\Omega)_{et}) = h^q$$

is an isomorphism for every geometric point $y : \operatorname{Spec}\Omega \to X_n$ of $X.$ (where the colimit is indexed by etale neighborhoods $V \to X_n$ of y), because

$h^q = h^q((\mathrm{Spec}\,k)_{et})$ is a direct summand of $\underset{\sim}{h}^q$. This is proved by comparing spectral sequences of type (16.5.1).

Finally, the canonical map $(X.',Y.') \to (X.\,,Y.)$ of the proof of Proposition 16.8 induces a map from the exact couple

$$\begin{array}{ccc} \bigoplus_{p,q} h^{p+q}((sk_pX.\,,sk_pY.)_{et}) & \longrightarrow & \bigoplus_{p,q} h^{p+q}((sk_{p-1}X.\,,sk_{p-1}Y.)_{et}) \\ & \nwarrow \qquad \swarrow & \\ & \bigoplus_{p,q} h^{p+q}((sk_pX.\,,sk_{p-1}X.\cup sk_pY.)_{et}) & \end{array} \tag{16.9.3}$$

to (16.8.2) which induces an isomorphism of spectral sequences starting at the E_2-level. Thus, (16.9.1) is isomorphic to the spectral sequence associated to the colimit of exact couples of type (16.9.3)′ for the pairs $(\Delta U..\,,\Delta\psi_{Y.}(U..))$ indexed by $U.. \in nHR(X.)$. The map (16.9.2) determines a map $\underset{\to}{\mathrm{colim}}$ (16.4.2)′ $\to$ $\underset{\to}{\mathrm{colim}}$ (16.9.3)′, and hence a map (16.5.1) $\to$ (16.9.1). By Proposition 15.2 this map of spectral sequences is an isomorphism as asserted. ∎

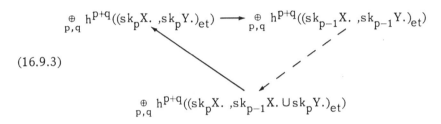

17. POINCARÉ DUALITY AND LOCALLY COMPACT HOMOLOGY

In this final chapter, we sketch a proof of Poincaré duality based upon the material of Chapters 14 and 15. Our proof is topological in nature and should be applicable to generalized cohomology theories.

We restrict our attention to (separated) algebraic varieties over a given algebraically closed field k although much of our discussion relativizes to more general base schemes. Poincaré duality (Theorem 17.6) is presented as an isomorphism between cohomology and "homology with locally compact supports." Such homology groups have been previously considered (for example, by R. Hartshorne in [49] and by S. Bloch and A. Ogus in [11]), defined in terms of cohomology by implicitly using duality. Our use of (relative) homology of etale homotopy types is both direct and natural, leading quickly to cycle classes and the duality map.

In Theorem 17.1, we reinterpret the cohomological purity theorem as a Thom isomorphism theorem. The fundamental homology class of a variety is constructed in Proposition 17.3, immediately permitting the construction of (homology) cycle classes. The relationship between these cycle classes and the Thom (cohomology) classes is described in Proposition 17.4. Theorem 17.6 asserts that cap product with the fundamental homology class of a smooth variety determines an isomorphism of the form required. As is apparent from our sketch of a proof of this theorem, this approach to duality is straightforward but dependent upon various (omitted) technical details concerning cap products.

In order to define fundamental classes and duality isomorphisms which are natural with respect to self maps of a variety X over k which are not k linear, we employ the *Tate twist* sending an abelian sheaf F on Et(X) to F(i) for $i \in \mathbf{Z}$. Namely, let m be a positive integer invertible in k and let $\mathbf{Z}/m(1)$ denote the (constant) sheaf on $Et(X)$ of m-th roots of unity with their prescribed behavior under $Aut(X)$. (Of course, an automorphism of X is not a map of $Et(X)$.) We define $F(i)$ for any sheaf F of $\mathbf{Z}/m$-modules by

$$F(i) = F \otimes \mathbf{Z}/m(i)$$

where $\mathbf{Z}/m(i)$ designates the i-th tensor product of $\mathbf{Z}/m(i)$ with itself for $i \geq 0$ and $\mathbf{Z}/m(i) = Hom(\mathbf{Z}/m(-i), \mathbf{Z}/m)$ for $i \leq 0$.

By an *algebraic variety* over k, we shall mean a reduced, irreducible scheme which is finite type and separated over k.

THEOREM 17.1. *Let X be a smooth algebraic variety over k, let $i: Z \to X$ be a smooth closed subvariety of codimension e, and let m be a positive integer invertible in k. For any locally constant, constructible sheaf M of $\mathbf{Z}/m$ modules on $Et(X)$, there is a canonical Thom isomorphism*

$$H^*(Z, i^*M) \longrightarrow H_Z^{*+2e}(X, M(e)) .$$

Moreover, if $\tau \in H_Z^{2e}(X, \mathbf{Z}/m(e))$ corresponds to $1 \in H^0(Z, \mathbf{Z}/m)$ under this isomorphism (τ is called the Thom class of $i: Z \to X$), then this isomorphism is given by cup product with τ using the pairing of Corollary 14.8:

$$H^*(T_{X/Z}, M) \otimes H^{2e}(T_{X/Z}, T_{X/Z} \underset{X}{\times} (X-Z); \mathbf{Z}/m(e))$$
$$\to H^{*+2e}(T_{X/Z}, T_{X/Z} \underset{X}{\times} (X-Z); M(e)) .$$

Proof. We employ the following spectral sequence

$$E_2^{p,q} = H^p(X, i_* R^q i^! M) \Longrightarrow H_Z^{p+q}(X, M) \tag{17.1.1}$$

obtained as the spectral sequence of the composite functor $H_Z^0(X, \ ) = H^0(X, \ ) \circ i_* \circ i^!$. By the *cohomological purity theorem* ([7], XVI.3.7 and XVI.3.8), each of the maps

$$(17.1.2)\qquad \begin{aligned} H^p(X,i_*i^*M)\otimes H^0(X,i_*R^{2e}i^!Z/m) &\to H^p(X,i_*R^{2e}i^!M) \\ &\to H_Z^{p+2e}(X,M) \end{aligned}$$

is an isomorphism. Because $R^{2e}i^!Z/m$ is canonically isomorphic to $Z/m(-e)$ on Z ([7], XVI.3.10), the asserted canonical isomorphism now follows.

The isomorphisms for any $F \in AbSh(X)$

$$H_Z^*(X,F) \simeq H^*(X,X-Z;F) \simeq H^*(T_{X/Z},T_{X/Z} \underset{X}{\times} (X-Z);F)$$

are given by Proposition 14.6 and (15.5.2). Using these isomorphisms and the isomorphism $H^*(Z,i^*F) \simeq H^*(T_{X/Z},F)$ of Proposition 15.2, we identify the composition of (17.1.2) as the asserted cup product by using the structure of (17.1.1) as a module over the corresponding spectral sequence with Z/m in place of M. ∎

One can view $\tau \in H_Z^{2e}(X,Z/m(e))$ as a (cohomology) cycle class for $i:Z \to X$. Such a definition may be extended to possibly singular Z by choosing any Zariski open X' of X containing $X-Z$ such that $Z' = Z \cap X'$ is smooth as well as dense in Z and observing that the natural restriction map

$$H_Z^{2e}(X,Z/m(e)) \to H_{Z'}^{2e}(X',Z/m(e)) \,.$$

is an isomorphism (cf. [59], VI.9.1).

In [49], R. Hartshorne considers (in the context of DeRham cohomology) the functor sending X to $H_X^{2N-*}(W)$ as a homology theory for X, where $X \to W$ is a closed immersion of a possibly singular complex algebraic variety X in a smooth variety W of dimension N. In particular, the fundamental homology class for X is the image of $1 \in H^0(X')$ for some smooth, dense $X' \subset X$ under the composition $H^0(X') \xrightarrow{\sim} H_{X'}^{2(N-n)}(W') \xrightarrow{\sim} H_X^{2(N-n)}(W)$ (as above).

Our approach to homology is to consider the homology of etale homotopy types.

PROPOSITION 17.2. *For any algebraic variety* X *over* k *and any constructible sheaf* F *on* Et(X), *we define the cohomology with compact supports,* $H_c^*(X,F)$, *to be*

$$H_c^*(X,F) = H^*(\overline{X}, j_! F) \simeq H^*(\overline{X}, Y; j_* F)$$

and the homology with locally compact supports, $H_*^{\ell.c.}(X,F^\vee)$, *to be*

$$H_*^{\ell.c.}(X,F^\vee) = H_*(\overline{X},(j_!F)^\vee) \simeq H_*(\overline{X},Y;(j_*F)^\vee)$$

where $j: X \to \overline{X}$ *is a compactification of* X *(i.e., a Zariski open inclusion with* $\overline{X}$ *a complete algebraic variety and* X *dense in* $\overline{X}$ *). Then* $H_c^*(X,F)$ *and* $H_*^{\ell.c.}(X,F^\vee)$ *are well defined, independent of the compactification up to canonical isomorphism, with* $H_*^{\ell.c.}(X,F^\vee)$ *naturally dual to* $H_c^*(X,F)$. *Moreover, a Zariski open immersion* $g: U \to X$ *and a proper map* $f: X \to Y$ *induce maps*

$$g_*: H_c^*(U, g^*F) \to H_c^*(X,F), \quad f^*: H_c^*(Y, f_*F) \to H_c^*(X,F) .$$

Proof. M. Nagata has proved the existence of a compactification ([60]). The duality of $H^*(\overline{X}, j_!F)$ and $H_*(\overline{X},(j_!F)^\vee)$ is given by Proposition 7.6, whereas Definitions 14.2 and 14.9 imply the isomorphisms

$$H^*(\overline{X},j_!F) \simeq H^*(\overline{X},Y;j_*F), \quad H_*(\overline{X},(j_!F)^\vee) \simeq H_*(\overline{X},Y;(j_*F)^\vee) .$$

Let $j: X \to \overline{X}$ and $j': X \to \overline{X}'$ be compactifications of X, and define $\overline{X}''$ to be the closure of $\Delta(X)$ in $\overline{X} \times \overline{X}'$ (where all products are to be viewed as products of varieties over Spec k). Because $\Delta(X)$ is closed (by separability) and dense in $(X \times X) \cap \overline{X}''$, X is isomorphic to $(X \times X) \cap \overline{X}''$ so that $j'': X \to \overline{X}''$ is another compactification. Because $\overline{X}$ and $\overline{X}'$ are separated, $\Delta(X)$ is closed in $\overline{X} \times X$ and $X \times \overline{X}'$, so that

$$(\overline{X} \times X) \cap \overline{X}'' = \Delta(X) = (X \times \overline{X}') \cap \overline{X}'' .$$

Thus, $pr_1 : \overline{X}'' \to \overline{X}$ and $pr_2 : \overline{X}'' \to \overline{X}'$ restrict to isomorphisms $j''(X) \xrightarrow{\sim} j(X)$, $j''(X) \to j'(X)$ and restrict to maps $\overline{X}'' - j''(X) \to \overline{X} - j(X)$, $\overline{X}'' - j''(X) \to \overline{X}' - j'(X)$.

Consequently, to check that $H_c^*(X,F)$ (and, thereby, $H_*^{\ell.c.}(X,F^\vee)$) is independent of $j : X \to \overline{X}$ up to canonical isomorphism, we may consider $j'' : X \to \overline{X}''$ mapping to $j : X \to \overline{X}$ via a proper map $\overline{f} : \overline{X}'' \to \overline{X}$ restricting to an isomorphism $j''(X) \xrightarrow{\sim} j(X)$ and a map $\overline{X}'' - j''(X) \to \overline{X} - j(X)$. The proper base change theorem ([59], VI.2.3) implies that $R^q\overline{f}_*(j''_!F) = 0$ for $q > 0$ and $\overline{f}_*(j''_!F) = j_!F$, so that $\overline{f}^* : H^*(\overline{X},j_!F) \to H^*(\overline{X}'',j''_!F)$ is an isomorphism as required.

The asserted functorality of $H_c^*(\ )$ with respect to a Zariski open immersion $U \to X$ is immediate from the observation that $(\overline{X},\overline{X}-X)$ maps to $(\overline{X},\overline{X}-U)$. Functorality for $f : X \to Y$ proper is deduced from the observation that if $j : X \to \overline{X}$ and $\overline{j} : Y \to \overline{Y}$ are compactifications, then as argued above f induces $\overline{f} : (\overline{X}',\overline{X}' - j'(X)) \to (\overline{Y},\overline{Y} - \overline{j}(Y))$ where $j' : X \to \overline{X}'$ is the compactification defined by $\Delta : X \to \overline{X} \times \overline{Y}$. ∎

The above proof that $X \mapsto H_*^{\ell.c.}(X,F^\vee)$ is independent of a choice of compactification is much simpler than Hartshorne's proof that $X \mapsto H_X^{2N-*}(W)$ is independent of a choice of embedding of X in a smooth complex variety W. For us, the duality of $H_*^{\ell.c.}(X,F^\vee)$ and $H_c^*(X,F)$ is merely formal and Poincaré duality asserts that $H_*^{\ell.c.}(X,Z/m(-n))$ is isomorphic to $H^{2n-*}(X,Z/m)$ whenever X is smooth. In Hartshorne's approach, the isomorphism between $H_X^{2N-*}(W)$ and $H^{2n-*}(X)$ for X smooth of dimension n is merely the Thom isomorphism theorem, whereas Poincaré-Lefschetz duality implies that $H_X^{2N-*}(W)$ is dual to $H_c^*(X)$.

For complex algebraic varieties, one may relate these two definitions of the homology of X with locally compact supports (namely, $H_X^{2N-*}(W)$ with X a closed subvariety of a smooth variety W of dimension N and $H_*(\overline{X},\overline{X}-X)$ with X a Zariski open of a complete variety $\overline{X}$) using Alexander duality. Using cap product with the fundamental homology class of X as in the proof of Theorem 17.6, one compares the long exact sequences

$$\cdots \to H_X^{2N-*}(W) \to H^{2N-*}(W) \to H^{2N-*}(W-X) \to \cdots$$

$$\cdots \to H_*(\overline{X},Y) \to H_*(\overline{W},Y) \to H_*(\overline{W},\overline{X}) \to \cdots$$

where $\overline{W}$ is a smooth compactification of W, $\overline{X} = \overline{W} - (W-X)$, and $Y = \overline{X} - X = \overline{W} - W$. The isomorphisms $H^{2N-*}(W) \simeq H_*(\overline{W},Y)$ and $H^{2N-*}(W-X) \simeq H_*(\overline{W},\overline{X})$ follow from Alexander duality for Y in $\overline{W}$ and X in $\overline{W}$.

The fact that $H_c^{2n}(X,Z/m(n)) \simeq Z/m$ is standard and the proof we give below is independent of etale homotopy theory. We provide a proof, nonetheless, because of its simplicity relative to its importance.

PROPOSITION 17.3. *Let* X *be an algebraic variety of dimension* n *over* k. *For* m *a positive integer invertible in* k, *there is a canonical isomorphism*

$$H_c^{2n}(X,Z/m(n)) \simeq Z/m .$$

The fundamental homology class $\chi_X \in H_{2n}^{\ell.c.}(X,Z/m(-n))$ *is defined to be the dual of* $1 \in Z/m \simeq H_c^{2n}(X,Z/m(n))$.

Proof. Because any proper closed subvariety of X has Z/m cohomological dimension $\leq 2n-2$, it suffices to replace X by any variety birationally equivalent to it. As argued in the proof of Theorem 11.6 (given in [7]), we may thus assume that there exists a compactification $j: X \to \overline{X}$, a proper map $\overline{f}: \overline{X} \to P^{n-1}$, and a Zariski open immersion $j': V \to P^{n-1}$ such that $f = \overline{f}_| : f^{-1}(V) = X \to V$ is proper and smooth with connected curves as fibres.

The Kummer exact sequence in AbSh(X)

$$0 \longrightarrow Z/m(1) \longrightarrow O^* \xrightarrow{m} O^* \longrightarrow 1$$

implies that $R^2 f_* Z/m(1)$ is canonically isomorphic to $Pic_{X/V}/m$, where $Pic_{X/V}$ is the relative Picard sheaf on V. Using the isomorphism

$\deg : Pic_{X/V}/m \xrightarrow{\sim} \mathbb{Z}/m$ (implied by the isomorphism of groups $\deg : \mathrm{Pic}(C)/m \xrightarrow{\sim} \mathbb{Z}/m$ for any smooth, complete, connected curve C over k), we conclude that

$$R^2\bar{f}_*(j_!\mathbb{Z}/m(1)) \simeq j_!'\mathbb{Z}/m .$$

Using the Leray spectral sequence, we conclude the isomorphisms

$$H^{2n}(\bar{X}, j_!\mathbb{Z}/m(n)) \xrightarrow{\sim} H^{2n-2}(P^{n-1}, R^2\bar{f}_*(j_!\mathbb{Z}/m(n)))$$
$$\xrightarrow{\sim} H^{2n-2}(P^{n-1}, j_!'\mathbb{Z}/m(n-1)) .$$

The proof is now completed by induction, the case $n = 1$ implied by the canonical isomorphisms

$$H_c^2(X, \mathbb{Z}/m(1)) \simeq H^2(\bar{X}, \mathbb{Z}/m(1)) \simeq \mathrm{Pic}(\bar{X})/m$$

for a smooth connected curve X with smooth compactification $j : X \to \bar{X}$. ∎

In the following proposition, we define the homology cycle class γ_Z of a closed immersion $i : Z \to X$ and we show that cap product with χ_X (the Poincaré duality map) sends the Thom class τ to γ_Z.

PROPOSITION 17.4. *Let X be a smooth algebraic variety of dimension n over k, let $i : Z \to X$ be the closed immersion of a closed subvariety Z of dimension d, and let m be an integer invertible in k. We define the homology cycle class γ_Z of $i : Z \to X$ by*

$$\gamma_Z = i_*(\chi_Z) \in H_{2d}^{\ell.c.}(X, \mathbb{Z}/m(-d)) .$$

There is a natural cap product pairing

$$\cap : H_Z^*(X, \mathbb{Z}/m(n-d)) \otimes H_{2n}^{\ell.c.}(X, \mathbb{Z}/m(-n)) \to H_{2n-*}^{\ell.c.}(Z, \mathbb{Z}/m(-d))$$

such that $\tau \cap \chi_X = \gamma_Z$.

Proof (Sketch). We consider the following maps

$$Z \xrightarrow{i} X \xrightarrow{j} \bar{X}, \quad Z \xrightarrow{j'} Y' \xrightarrow{i'} \bar{X}$$

where j is a compactification of X, i' is the closed immersion of $Y' = \bar{X} - (X-Z)$ in $\bar{X}$, and j' is the associated Zariski open immersion. The proof of Proposition 17.3 implies the isomorphism $H_c^*(Z,F) \simeq H^*(Y',Y; j_*F)$ even if Z is not dense in Y', where $Y = Y' - Z$. The cap product pairing is obtained by applying Proposition 14.11 to

$$T_{\bar{X}/Y'} \underset{\bar{X}}{\times} (X-Z) \cup T_{\bar{X}/Y} \underset{\bar{X}}{\times} X \to T_{\bar{X}/Y'} \underset{\bar{X}}{\times} X$$

and using the excision isomorphism of Proposition 15.5.

As in the discussion following 17.1, we may assume Z is a smooth subvariety of X. The exactness of $j_!$, $j'_!$ and the equality

$$i'^! \circ j_! = j'_! \circ i^!$$

imply the isomorphism

$$R^q i'^!(j_! F) \simeq j'_!(R^q i^! F)$$

for any $q \geq 0$ and $F \in \mathrm{AbSh}(X)$. The spectral sequence for the composite functor $H^0(\bar{X}, \ ) \circ i'_* \circ i'^!$ applied to the sheaf $F = j_! \mathbf{Z}/m(n)$ and our knowledge of $R^q i^! \mathbf{Z}/m$ (see the proof of Proposition 17.1) imply the natural isomorphism

$$H^{2n}_{Y'}(\bar{X}, j_! \mathbf{Z}/m(n)) \simeq H^{2d}(\bar{X}, i'_* j'_! R^{2n-2d} i^! \mathbf{Z}/m(n)) .$$

Since $R^{2n-2d} i^! \mathbf{Z}/m(n-d) \simeq \mathbf{Z}/m$ on $Et(Z)$ (as seen in Proposition 17.1), the cup product pairing

(17.4.1)
$$H^{2d}(\bar{X}, i'_* j'_! \mathbf{Z}/m(d)) \otimes H^0(X, i_* R^{2n-2d} i^! \mathbf{Z}/m(n-d)) \to H^{2d}(\bar{X}, i'_* j'_! R^{2n-2d} i^! \mathbf{Z}/m(n))$$

is an isomorphism. We conclude that for some unit $c \in \mathbf{Z}/m$

$$(17.4.2) \quad \chi_X^{\vee} = c(\gamma_Z^{\vee} \cup \tau) \in H_Y^{2n}(\bar{X}, j_! \mathbf{Z}/m(n)) \simeq H^{2n}(\bar{X}, j_! \mathbf{Z}/m(n)) ,$$

where $\chi_X^{\vee}$ and $\gamma_Z^{\vee}$ are the dual cohomology classes to χ_X and γ_Z.

To verify that $c = 1$ in (17.4.1) (this is not necessary for the proof of Theorem 17.6), one employs the naturality of the above cup product pairing and the fact that $Z \to X$ is locally (in the etale topology) isomorphic to $\mathbf{A}^d \to \mathbf{A}^n$ (cf. [7], XVI.3). This reduces the determination of c to the familiar case in which $Z \to X$ is the linear closed immersion $\mathbf{P}^d \to \mathbf{P}^n$.

Finally, the equality $\tau \cap \chi_X = \gamma_Z$ is merely the dual to the equality $\chi_X^{\vee} = \gamma_Z^{\vee} \cup \tau$ proved above. ∎

In order to consider locally constant (rather than only constant) abelian sheaves in our Poincaré duality theorem, we require the following proposition.

PROPOSITION 17.5. *Let* $f: X' \to X$ *be a finite, etale map of degree* e *of algebraic varieties over* k. *For any abelian sheaf* F *on* $\mathrm{Et}(X)$, *there exists a natural map*

$$\mathrm{tr} : f_* f^* F \to F \quad \textit{in} \quad \mathrm{AbSh}(X)$$

such that composition with the canonical map $F \to f_* f^* F$ *is multiplication by* e. *This map induces a transfer map*

$$\mathrm{tr}^* : H_*^{\ell.c.}(X, F^{\vee}) \to H_*^{\ell.c.}(X', (f^*F)^{\vee})$$

whenever F *is constructible. Moreover, for any positive integer* m *invertible in* k, $\mathrm{tr}^*(\chi_X) = \chi_{X'} \in H_{2n}^{\ell.c.}(X', \mathbf{Z}/m(-n))$ *where* $n = \dim(X)$, *providing* e *is relatively prime to* m.

Proof. If $U \to X$ in $\mathrm{Et}(X)$ is such that $X' \times_X U \simeq \amalg U$ over U, then we define

$$\mathrm{tr} : f_* f^* F(U) \xrightarrow{\sim} f^* F(\amalg U) \xrightarrow{\sim} \oplus F(U) \longrightarrow F(U)$$

to be the sum map. This determines $\mathrm{tr}: f_*f^*F \to F$. Whenever $X' \times_X U \simeq \amalg U$, $F(U) \to f_*f^*F(U)$ equals the diagonal $F(U) \to \oplus F(U)$; thus, the composition $F \to f_*f^*F \to F$ equals multiplication by e.

We exhibit $\mathrm{tr}_*: H_c^*(X', f^*F) \to H_c^*(X,F)$ dual to tr^*. As in the proof of Proposition 17.3, choose compactifications $j: X \to \bar{X}$ and $j': X' \to \bar{X}'$ fitting in the following commutative diagram

$$\begin{array}{ccccc} X' & \xrightarrow{j'} & \bar{X}' & \longleftarrow & \bar{X}' - X' \\ \downarrow f & & \downarrow \bar{f} & & \downarrow \\ X & \xrightarrow{j} & \bar{X} & \longleftarrow & \bar{X} - X \end{array} .$$

Then $\mathrm{tr}: f_*f^*F \to F$ determines

$$\mathrm{tr}: \bar{f}_* j'_! f^*F \simeq j_! f_* f^*F \to j_! F$$

and hence the map tr_* defined as the composition

$$H_c^*(X', f^*F) = H^*(\bar{X}', j'_! f^*F) \simeq H^*(\bar{X}, \bar{f}_* j'_! f^*F)$$
$$\to H^*(\bar{X}, j_!F) = H_c^*(X,F) .$$

We prove the equality $f^*(\chi_X^{\vee}) = e\chi_{X'}^{\vee} \in H_c^{2n}(X', \mathbf{Z}/m(n))$ which immediately implies the equality $\mathrm{tr}^*(\chi_X) = \chi_{X'} \in H_{2n}^{\ell.c.}(X', \mathbf{Z}/m(-n))$ providing e is relatively prime to m. As in the proof of Proposition 17.3, we may assume the existence of a proper map $\bar{g}: \bar{X} \to \mathbf{P}^{n-1}$ and a non-empty Zariski open $\bar{j}: V \to \mathbf{P}^{n-1}$ such that $\bar{g}$ restricted to $X = \bar{g}^{-1}(V)$ is proper and smooth with connected fibres. Then

$$\bar{f}^*: H^{2n}(\bar{X}, j_! \mathbf{Z}/m(n)) \simeq H^{2n-2}(\mathbf{P}^{n-1}, \bar{j}_! R^2 g_* \mathbf{Z}/m(n))$$
$$\to H^{2n-2}(\mathbf{P}^{n-1}, \bar{j}_! R^2 (g \circ f)_* \mathbf{Z}/m(n)) \simeq H^{2n}(\bar{X}', j'_! \mathbf{Z}/m(n))$$

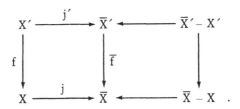

is induced by $f^*: R^2 g_* \mathbf{Z}/m(1) \to R^2(g \circ f)_* \mathbf{Z}/m(1)$. We identify this map (as in the proof of Proposition 17.3) with

$$f^* : Pic_{X/V}/m \to Pic_{X'/V}/m \, .$$

Since this map sends a divisor to its inverse image, it multiplies degrees by e and hence is identified with multiplication by e on $\mathbf{Z}/m$ as required. ■

We conclude with the following sketch of Poincaré duality (see also [7], XVIII or [59], VI.11.1).

THEOREM 17.6. *Let* X *be a smooth algebraic variety of dimension* n *over* k *and let* m *be a positive integer invertible in* k. *Then the duality map given by cap product with* $\chi_X \in H_{2n}^{\ell.c.}(X,\mathbf{Z}/m(-n))$,

$$D_X : H^i(X,M) \to H_{2n-i}^{\ell.c.}(X,M(n)^\vee) \, ,$$

is an isomorphism for all $i \geq 0$ *whenever* M *is a locally constant, constructible sheaf of* $\mathbf{Z}/m$ *modules on* Et(X).

Proof (Sketch). The proof proceeds by induction on $n = \dim(X)$.

Stage 1. The definition of D_X. Let $j : X \to \overline{X}$ be a compactification with reduced complement $Y = \overline{X} - X$. Using (15.5.1), we identify $H_*^{\ell.c.}(X,M^\vee)$ with $H_*((T_{\overline{X}/\overline{X}} \underset{\overline{X}}{\times} X)_{et}, (T_{\overline{X}/Y} \underset{\overline{X}}{\times} X)_{et}; M^\circ)$, so that Proposition 14.11 for $\phi \to T_{\overline{X}/\overline{X}} \underset{\overline{X}}{\times} X$, $T_{\overline{X}/Y} \underset{\overline{X}}{\times} X \to T_{\overline{X}/\overline{X}} \underset{\overline{X}}{\times} X$ provides the cap product pairing

$$D_X : H^i(X,M) \otimes H_{2n}^{\ell.c.}(X,\mathbf{Z}/m(-n)) \to H_{2n-i}^{\ell.c.}(X,M(n)^\vee) \, .$$

Stage 2. Reduction to $M = \mathbf{Z}/q$, q a prime dividing m. Since M is a direct sum of its q-primary components, we may assume M is q-primary. Fix a closed point x of X, let G denote the image of $\pi_1(X,x)$ in $\mathrm{Aut}(M_x)$, and let $G_q \subset G$ denote the q-Sylow subgroup of G. Let

$g: X',x' \to X,x$ be the finite etale covering of degree prime to q associated to the inverse image of G_q, so that $\pi_1(X',x')$ acts on g^*M through G_q.

Because G_q and g^*M are q-primary, g^*M admits a filtration whose successive quotients are $\mathbf{Z}/q$. Hence, if $D_{X'}$ is an isomorphism for $\mathbf{Z}/q$ as coefficients, then $D_{X'}$ is also an isomorphism for g^*M as coefficients. By Proposition 17.5, the following squares sign-commute

$$\begin{array}{ccccc} H^i(X,M) & \xrightarrow{g^*} & H^i(X',g^*M) & \xrightarrow{tr_*} & H^i(X,M) \\ \downarrow{\scriptstyle D_X} & & \downarrow{\scriptstyle D_{X'}} & & \downarrow{\scriptstyle D_X} \\ H^{\ell.c.}_{2n-i}(X,M(n)^\vee) & \xrightarrow{tr^*} & H^{\ell.c.}_{2n-i}(X',g^*M(n)^\vee) & \xrightarrow{g_*} & H^{\ell.c.}_{2n-i}(X,M(n)^\vee) \end{array}$$

where $tr_*: H^i(X',g^*M) \to H^i(X,M)$ is induced by $g_*g^*M \to M$. Because the degree of g is prime to q and M is q-primary, g^* is injective and g_* is surjective. Thus, D_X is an isomorphism whenever $D_{X'}$ is.

Stage 3. Reduction to a base of Zariski opens of X. If V and V' are Zariski open subvarieties of X with $U = V \cup V'$ and $W = V \cap V'$, the duality maps fit in a map of long exact sequences

$$\begin{array}{ccccccc} \cdots \to & H^i(U,\mathbf{Z}/m) & \to & H^i(V,\mathbf{Z}/m) \oplus H^i(V',\mathbf{Z}/m) & \to & H^i(W,\mathbf{Z}/m) & \to \cdots \\ & \downarrow{\scriptstyle D_U} & & \downarrow{\scriptstyle D_V \oplus D_{V'}} & & \downarrow{\scriptstyle D_W} & \\ \cdots \to & H^{\ell.c.}_{2n-i}(U,\mathbf{Z}/m(-n)) & \to & H^{\ell.c.}_{2n-i}(V,\mathbf{Z}/m(-n)) \oplus H^{\ell.c.}_{2n-i}(V',\mathbf{Z}/m(-n)) & \to & H^{\ell.c.}_{2n-i}(W,\mathbf{Z}/m(-n)) & \to \cdots . \end{array} \tag{17.6.1}$$

Applying the 5-Lemma to (17.6.1), we conclude that D_V and $D_{V'}$ being isomorphic for $i-1$, i and D_W being injective for $i-1$ implies that D_U is injective for i. Thus, we may inductively conclude for all Zariski opens U of X that D_U is an isomorphism for $j < i$ and injective for i

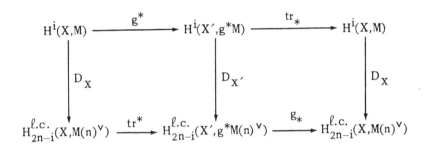

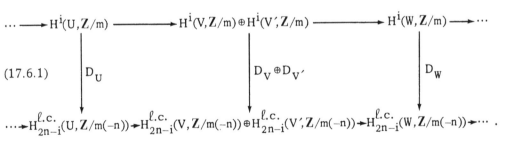

provided that D_V is an isomorphism for all i and all V in some base of Zariski opens. Similarly, D_U is surjective for i provided that D_V and $D_{V'}$ are isomorphisms for i and D_W is an isomorphism for $i-1$ and injective for i.

Stage 4. Reduction to $f: X \to V$ proper, smooth with connected curves as fibres and f compactifying to $\overline{f}: \overline{X} \to P^{n-1}$. As verified in the proof of Theorem 11.6 given in [7], X has a base of Zariski open subvarieties U' which are of the form $U' = U - Z$ where U admits a map $f: U \to V$ as above and f restricted to Z is finite, etale. By Stage 3 and induction, to verify the asserted reduction it suffices to prove that D_U is an isomorphism whenever $D_{U'}$ and D_Z are isomorphisms. This follows from the map of long exact sequences

$$
(17.6.2)\qquad
\begin{array}{ccccccccc}
\cdots \longrightarrow & H^i_Z(U,Z/m) & \longrightarrow & H^i(U,Z/m) & \longrightarrow & H^i(U',Z/m) & \longrightarrow \cdots \\
& \Big\downarrow & & \Big\downarrow{\scriptstyle D_U} & & \Big\downarrow{\scriptstyle D_{U'}} & \\
\cdots \longrightarrow & H^{\ell.c.}_{2n-i}(Z,Z/m(-n)) & \longrightarrow & H^{\ell.c.}_{2n-i}(U,Z/m(-n)) & \longrightarrow & H^{\ell.c.}_{2n-i}(U',Z/m(-n)) & \longrightarrow \cdots
\end{array}
$$

where the left vertical arrow is the composition of D_Z with the inverse of the Thom isomorphism of Proposition 17.1, and where the bottom sequence arises from the triple $\overline{X} - U \subset \overline{X} - U' \subset \overline{X}$ and the identification $H^{\ell.c.}_*(Z,F) \simeq H_*(\overline{X} - U', \overline{X} - U; F)$. The proof of the commutativity of (17.6.2) is based upon Proposition 17.4 which asserts that the maps $H^*_Z(U,Z/m) \to H^{\ell.c.}_{2n-*}(Z,Z/m(-n))$ of (17.6.2) are given by cap product with χ_X.

Stage 5. The case $n = 1$. This can be proved by lifting X to characteristic 0 and using Poincaré duality for Riemann surfaces. For X complete, the lifting $X_R \to \operatorname{Spec} R$ is proper and smooth, where R denotes the Witt vectors of k, so that Theorem 8.4, Proposition 8.6, and Proposition 8.7 apply. For X not complete, then the lifting $X_R \to \operatorname{Spec} R$ can be chosen to be an elementary fibration so that the arguments of Chapter 11 apply.

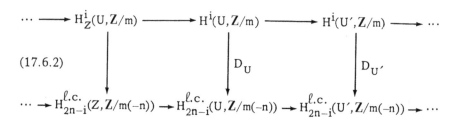

Stage 6. Inductive step. Let $f: X \to V$ be as asserted in Stage 4, with geometric fibre $F \to X$. By Stage 5 and induction, both D_F and D_V are isomorphisms. The proof is completed by verifying that cap product with χ_X determines a map from the Serre spectral sequence in cohomology for f_{et} (see Corollary 10.8) to the Serre spectral sequence in homology for the pair $(\bar{f}_{et}, g_{et})$ $(g = \bar{f}|: \bar{X} - X \to P^{n-1} - V)$ which is an isomorphism on E_2-levels (cf. [45]). ■

REFERENCES

[1] J.F. Adams: On the groups J(X)-I. Topology 2 (1963), 181-195.

[2] ―――――: On the groups J(X)-IV. Topology 5 (1966), 21-71.

[3] ―――――: *Stable Homotopy and Generalized Homology*. University of Chicago Press, 1972.

[4] J.F. Adams and Z. Mahmud: Maps between classifying spaces. Inventiones Math. 35(1976), 1-42.

[5] M. Artin: *Grothendieck Topologies*. Harvard Seminar Notes, 1972.

[6] ―――――: On the joins of hensel rings. Advances in Math. 7(1971), 282-296.

[7] M. Artin, A. Grothendieck, and J.-L. Verdier: SGA 4, *Theorie des Topos et Cohomologie Etale des Schemas*. Lecture Notes in Math. 269(1972), 270(1972), 305(1973), Springer.

[8] M. Artin and B. Mazur: *Etale Homotopy*. Lecture Notes in Math. 100, Springer, 1969.

[9] M.F. Atiyah: Thom complexes. Proc. Lon. Math. Soc. 11(1961), 291-310.

[10] P. Baum, W. Fulton, and R. MacPherson: Riemann Roch for singular varieties. Pub. I.H.E.S. 45(1975), 101-146.

[11] S. Bloch and A. Ogus: Gersten's conjecture and the homology of schemes. Ann. Scient. Ec. Norm. Sup., t. 7(1974), 181-202.

[12] A.K. Bousfield and E.M. Friedlander: Homotopy theory of Γ-spaces, spectra, and bisimplicial sets. Lecture Notes in Math. 658, Springer, (1978), 80-131.

[13] A.K. Bousfield and D.M. Kan: *Homotopy Limits, Completions, and Localizations*. Lecture Notes in Math. 304, Springer, 1972.

[14] G.E. Bredon: *Sheaf Theory*. McGraw-Hill, 1967.

[15] W. Browder: *Surgery on Simply Connected Manifolds*. Springer, 1972.

[16] K.S. Brown and S.M. Gersten: Algebraic K-theory as generalized sheaf cohomology. Lecture Notes in Math. 341, Springer (1973), 266-293.

[17] H. Cartan and S. Eilenberg: *Homological Algebra*. Princeton Univ. Press, 1956.

[18] D. Cox: Algebraic tubular neighborhoods II. Math. Scand. 42 (1978), 229-242.

[19] ________: Homotopy theory of simplicial schemes. Composito Mathematica 39 (1979), 263-295.

[20] ________: The etale homotopy type of varieties over R. Proc. A.M.S. 76 (1979), 17-22.

[21] ________: Spherical fibrations in algebraic geometry. Illinois J. Math. 24 (1980), 18-47.

[22] P. Deligne: Conjecture de Weil I. Publ. Math. I.H.E.S. 43 (1974), 273-307.

[23] ________: Theorie de Hodge III. Publ. Math. I.H.E.S. 44 (1974), 6-77.

[24] ________: *Cohomologie Etale* (SGA 4½). Lecture Notes in Math. 569, Springer, 1977.

[25] P. Deligne and D. Sullivan: Fibres vectoriels complexes a groupe structural discret. C. R. Acad. Sc. Paris t. 281 (1975), Serie A 1081-1083.

[26] A. Dold and D. Puppe: Homologie nicht additives functoren. Andwendugen. Ann. Inst. Fourier, Grenoble 11 (1961), 201-312.

[27] W. Dwyer and E. M. Friedlander: Etale K-theory and arithmetic. Bull. A.M.S., vol. 6, no. 3 (1982), 453-455.

[28] D. A. Edwards and H. M. Hastings: *Čech and Steenrod Homotopy Theory with Applications to Geometric Topology*. Lecture Notes in Math. 542, Springer, 1976.

[29] E. M. Friedlander: Fibrations in etale homotopy theory. Pub. Math. I.H.E.S. 42 (1972), 5-46.

[30] ________: $K(\pi,1)$'s in characteristic $p>0$. Topology 12 (1973), 9-18.

[31] ________: The etale homotopy theory of a geometric fibration. Manuscripta Mathematica 19 (1973), 209-244.

[32] ________: Unstable K-theories of the algebraic closure of a finite field. Comment. Math. Helvetici 50 (1975), 145-154.

[33] ________: Exceptional isogenies and the classifying spaces of simple Lie groups. Annals of Math. 101 (1975), 510-520.

[34] ________: Computations of K-theories of finite fields. Topology 15 (1976), 87-109.

[35] ________: Homological stability for classical groups over finite fields. Lecture Notes in Math. 551, Springer (1976), 290-303.

[36] ________: Maps between localized homogeneous spaces. Topology 16 (1977), 205-216.

[37] E. M. Friedlander: The infinite loop Adams conjecture via classification theorems for $\mathcal{F}$-spaces. Math. Proc. Camb. Phil. Soc. (1980), 109-150.

[38] ————: Etale K-theory I: Connections with etale cohomology and algebraic vector bundles. Inventiones math. 60 (1980), 105-134.

[39] ————: Etale K-theory II: Connections with algebraic K-theory. To appear in Ann. Scient. Ec. Norm. Sup.

[40] E. M. Friedlander and B. Parshall: Etale cohomology of reductive groups. Lecture Notes in Math. 854, Springer (1981), 127-140.

[41] E. M. Friedlander and S. Priddy: Karoubi's conjecture for finite fields. J. of Pure and Appl. Algebra 10 (1977), 233-238.

[42] P. Gabriel and M. Zisman: *Calculus of Fractions and Homotopy Theory.* Springer, 1967.

[43] S. M. Gersten: Higher K-theory of rings. Lecture Notes in Math. 341, Springer (1973), 3-42.

[44] J. Giraud: *Technique de Descente.* Bull. Soc. Math. France, Memoire No. 2, 1964.

[45] D. Gottlieb: Poincaré duality and fibrations. Proc. A.M.S. 76 (1979), 148-150.

[46] A. Grothendieck: Sur quelques points d' algebre homologique. Tohoku Math. Journ. IX (1957), 119-220.

[47] ————: *Elements de Geometrie Algebrique I and III.* Publ. Math. I.H.E.S. 4 (1960) and 11 (1961).

[48] ————: *Revetements Etale et Groupe Fundamental* (SGA 1). Lecture Notes in Math. 224, Springer, 1971.

[49] R. Hartshorne: On the DeRham cohomology of algebraic varieties. Pub. Math. I.H.E.S. 45 (1975), 5-99.

[50] ————: *Algebraic Geometry.* Springer, 1977.

[51] R. Hoobler and D. L. Rector: Arithmetic K-theory. Lecture Notes in Math. 418, Springer (1974), 78-95.

[52] S. Kleinerman: Thesis, Northwestern University, 1980.

[53] S. Lang: Algebraic groups over finite fields. Amer. J. Math. 78 (1956), 555-563.

[54] S. Lubkin: On a conjecture of Andre Weil. Amer. J. Math. 89 (1967), 443-548.

[55] S. MacLane: *Homology.* Springer, 1963.

[56] I. Madsen: On the action of the Dyer-Lashof algebra in $H_*(G)$. Pac. J. of Math. 60 (1975), 235-275.

[57] J. P. May: *Simplicial Objects in Algebraic Topology.* Van Nostrand, 1967.

[58] ________: *Classifying Spaces and Fibrations.* Memoirs of the A.M.S. 155 (1975).

[59] J. S. Milne: *Etale Cohomology.* Princeton Univ. Press, 1980.

[60] M. Nagata: Imbedding of an abstract variety in a complete variety. J. Math. Kyoto Univ., 2-1 (1962), 1-10.

[61] D. Quillen: Some remarks on etale homotopy and a conjecture of Adams. Topology 7 (1968), 111-116.

[62] ________: Cohomology of groups. Actes Congres Intern. Math. 1970, Tome 2, 47-51.

[63] ________: The Adams conjecture. Topology 10 (1971), 67-80.

[64] ________: On the cohomology and K-theory of the general linear group over finite fields. Annals of Math. 96 (1972), 552-586.

[65] ________: Higher algebraic K-theory I. Lecture Notes in Math. 341, Springer (1973), 85-147.

[66] J.-P. Serre: *Cohomologie Galoisienne.* Lecture Notes in Math. 5, Springer, 1964.

[67] C. Soulé: K-theorie des anneaux d'entiers de corps de nombres et cohomologie etale. Inventiones Math. 55 (1979), 251-295.

[68] R. Steinberg: Endomorphisms of linear algebraic groups. Memoirs of the A.M.S. 80 (1968).

[69] D. Sullivan: Genetics of homotopy theory and the Adams conjecture. Annals of Math. 100 (1974), 1-79.

[70] R. Thomason: Algebraic K-theory and etale cohomology. Preprint.

[71] C. Wilkerson: Self maps of classifying spaces. Lecture notes in Math. 418, Springer (1974), 150-157.

[72] Z. Wojtkowiak: On the action of $Gal(\tilde{Q},Q)$ on $BU(n)^{\wedge}$. Preprint.

[73] W. Dwyer, E. M. Friedlander, V. Snaith, and R. Thomason: Topological K-theory eventually surjects onto algebraic K-theory. To appear in Inventiones math.

INDEX

Library of Congress Cataloging in Publication Data

Friedlander, E. M. (Eric M.), 1944-
Etale homotopy of simplicial schemes.

(Annals of mathematics studies ; 104)
1. Homotopy theory. 2. Schemes (Algebraic geometry).
3. Homology theory. I. Title. II. Series.
QA612.3.F74 514'.24 81-47129
ISBN 0-691-08317-7 AACR2

Eric M. Friedlander is Professor of Mathematics at Northwestern University.

www.ingramcontent.com/pod-product-compliance
Ingram Content Group UK Ltd.
Pitfield, Milton Keynes, MK11 3LW, UK
UKHW042228130125
453571UK00001B/61

9 780691 083179